fire

This book is dedicated to
the keepers of the flame;
those who keep the fire burning.
Thank you!

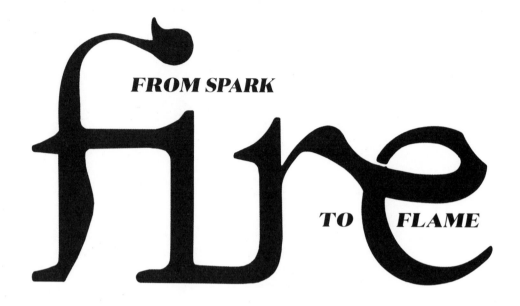

FROM SPARK fire TO FLAME

THE SCANDINAVIAN ART OF FIRE-MAKING

ØIVIND BERG

CARLTON
BOOKS

Contents

SPARK

I light a match and watch as a spark turns to flame on the piece of bark I used as tinder. I smell smoke as the spruce twigs crackle and pop. I stretch my cold hands towards the flames and feel the warmth of the fire slowly permeate my clothes to reach my skin and spread throughout my body. The light of the fire makes the darkness around my campfire even darker.

The fire forms a nucleus of heat; I am aware of the light fending off the darkness, keeping it at bay. The flames dance, coming to life, pulsating from the red-hot nucleus of the fire. The wood hisses as its vapours bubble to the surface and evaporate. The wood crackles, exploding as white and yellow sparks rising to the darkness above.

My iron kettle

I remember my first excursions with mother and father, always with their rituals. My father wore breeches, long stockings and heavy boots – summer or winter. His grey army rucksack always contained kindling, some matches, a time-worn coffee kettle packed in old newspapers and a large pocket knife – always the same gear.

He would lay a circle of stones on the ground and tear up some old newspaper as tinder, carefully placing his kindling on top of the paper in the shape of a fan. He would use his body to block the wind as he lit a match. It never failed; his fire always ignited with the first match.

As soon as the fire was burning he would gather more wood, fill his kettle with water and balance it perfectly on a stone near the flames. It never took much time for the water to boil. He would fill his kettle with exactly as much coffee as was necessary and let it boil no more than was needed, before removing the kettle from the flames and setting it aside. He would stir the coffee with a birch twig after it was brewed. My mother always turned her nose up at this ritual. She preferred her coffee without the taste of birch, but my father would simply smirk, saying coffee from a kettle should not taste like coffee at home. It should taste of the mountain and be as strong as dynamite. That was how he wanted it.

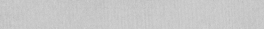

FIRE

The wheel was invented. Fire was mastered.
Early man learned to use fire when the need arose.
As time went by, the warmth and light of the fire
pit became a natural gathering place.

A raging fire is a dangerous thing. Man had to learn to use fire correctly. Mankind's control over fire eventually became his most powerful tool. The mastery of fire was the first clear distinction between man and animal.

This is said to have occurred about half a million years ago; perhaps the story goes something like this. A bolt of lightning started a fire. Hunters found a burnt animal carcass that had fallen victim to the flames. Perhaps they also discovered that meat was easier to chew when burnt, which led man to conserve meat and fish using heat.

Hunters learned to control wildfires in order to drive their prey to an area where it was easy to kill with a bow and arrow or a spear. Man learned to use wind and fire as a clever hunting technique.

Archaeological sites with charred earth in Africa and Eurasia are thought to be proof that our hominid ancestors, *Homo erectus* (upright man), used fire in a controlled manner. Archaeologists believe that *Homo erectus* lived between 40,000 and 1.8 million years ago. They also think that this species may be a common ancestor of modern man (*Homo sapiens*). Most anthropologists agree that *Homo erectus* was a hunting species that migrated from the African continent after learning to use tools and master fire. In the beginning, these men may have brought burning embers with them, having learned that embers stayed hot even after travelling long distances. Man eventually learned how to use tools to make fire.

Sites with charred earth and animal bones are considered proof that humans had finally mastered fire and changed their eating habits. The anthropologist Claude Lévi-Strauss argued that the transition from raw food to roasting or boiling our meals represents the first clear distinction between mankind and other animals. Our chances of survival and basic living conditions improved radically when fire became a source of heat and light and also a way of preparing meat and fish. Research indicates that fire hastened human migration from the savannahs of Africa towards temporary and permanent settlements and dwellings. The fire pit became "the hearth" around which human society would grow.

The hearth is no longer an integral part of our homes, but most people still feel a strong connection to dancing flames, and we are drawn to the fireside more for pleasure than heat.

Fire

Fire is visible light and infrared radiation from rapid combustion at a high temperature.

Flames result from a chemical reaction between oxygen in the air and a combustible substance. This process will only start when an increase in temperature causes a combustible substance to turn to gas and ignite. Tiny carbon particles then become so hot that they emit light. The light we see is flames, and this is what we call fire.

God gave us fire of glowing flame.
With caution, our companion;
Friendly flames give warmth and light.
We are grateful, oh fire!
We promise to tend you with care.

Gisle Skeie (2007)

Prometheus, God of Fire

In Greek mythology, Prometheus was the son of the Titan Iapetus. When Man's arrogance led us to disobey the gods and refuse to offer worship, Zeus decided to punish mankind by taking fire away from us. Prometheus stole fire from the gods and returned it to mankind.

Fire as a symbol

Fire has also been of great importance to religion and ritual. Fires are lit, they grow and expire; they are a symbol of life itself – and of love. Fire has its dark side too – intrinsic danger and destruction, symbolizing pain and loss.

Fire has not lost its symbolic significance for modern society. We light candles in celebration and worship, and as a sign of sorrow. The Scandinavian countries have always used a bonfire to solemnize the turning of the sun. We still use torches in protest marches and at gatherings to show support for causes we believe in.

The Olympic flame is carried in the hands of relay runners from its home in Olympia, Greece, to the location of the Games. The flame is used to light the Olympic cauldron, which is not extinguished until the Games have ended. The lighting and extinguishing of the cauldron are important symbolic events. The Olympic torch has been burning at Morgedal since the 1952 Winter Olympics, as it does in Lillehammer following the 17th Winter Olympics in 1994. These fires are similar to the eternal flames found at temples in India and Azerbaijan.

Fire is an essential part of mythology and has played an important role in religion throughout history. The Bible tells of a pillar of flame that guided the Israelites through the desert at night, away from the Pharaoh's army:

And the LORD went before them by day in a pillar of a cloud, to lead the way; and by night in a pillar of fire, to give them light, so they could travel day and night.

The Greek philosophers were interested in the physical properties of fire. Aristotle's theory that all substances are based on four elements (earth, water, air and fire) was considered a scientific truth for centuries. The theory proposed that wood consisted of earth and water. A tree has its roots planted in the wet earth, takes air from the sky, and absorbs the fire of the sun. Trees contain all four elements: earth, water, air and fire. Trees can only be used as fuel when dried, after releasing the water element. Fire and air are released as wood burns, leaving only earth in the form of ashes. This sounded logical to anyone who had ever tended a fire, until a doctor called Georg Stahl proposed a new theory in the 1700s. According to Stahl, all substances that can burn release the same element. Stahl called that element *phlogiston*. He argued that sulphur was almost pure phlogiston because it leaves almost nothing behind when it burns. He asserted that a candle burning inside a closed container would extinguish when the air could no longer hold more phlogiston. Today we know the opposite to be true; a flame dies when the oxygen in the air has gone. This discovery was made by Antoine Lavoisier, but not until the end of the 1700s.

FIRE

And Fire said to the wayfarer:

You appear to be freezing,

you enshroud your body

to protect your homeless heart.

Yet I shall light your way to

a gateway of ashes.

Are you prepared

to lay bare your naked body

and approach

a countenance of light?

Hans Børli, Glåmdalen, Norway, 29 July 1961

CAMPFIRES

*The campfire is an important
part of outdoor life.
Campfires provide warmth and fellowship.
A warm kettle of coffee beside a
campfire is mandatory.
One seldom sees discord or hostility
in the orange glow of a campfire.*

The words used by the Sami people for lighting a campfire, *dålle bïejem*, mean to gather around it. There are several words to describe a campfire, such as *dållå* in the Sami language, depending on what kind of campfire you mean: its size, how it is built, what kind of firewood you use and the ground on which it is built. This is not incidental; it is the result of hundreds of years of knowledge and experience.

It shows us how important fire was for indigenous people. A campfire invites fellowship; conversations flow effortlessly – even among strangers. First Nation Tribes in America would gather around a campfire to discuss alliances and share the peace pipe to confirm their agreements.

One can also sit beside a campfire in silence – a comforting silence in which conversation is not necessary. One feels welcome and united in fellowship. If I am alone in the woods, I often sit beside my campfire and feel that I belong to something much greater than myself. Fire can be a great travelling companion, bringing us closer to nature. When I am tired, cold and wet, a fire dries my clothes and keeps me warm, even on the coldest nights.

Fire pit, fuel and flame

Lighting a fire is easier said than done. You must prepare properly to get a fire lit and keep it burning. There are four main decisions to make: where should I build my fire? How should I build it? What tinder and kindling do I need to get it burning? And what fuel will keep it burning?

The fire pit

A good location needs to be sheltered from wind to maximize heat and keep a fire burning slowly. Even though many people prefer to build a campfire beside a rock wall or behind a large boulder, I prefer my fires in the open. A sheltered campfire does have its advantages. It burns more slowly and heat is reflected from the rock wall or boulder. In my experience, however, the disadvantages outweigh the advantages.

Wind moving around a rock creates an eddy that sends smoke swirling near the fire. I do not like smoke in my eyes. That is why I build my fires beneath the heavens. The slightest wind will blow the smoke away and, if the wind direction changes, you can simply move and sit somewhere else. On windy days I set up camp near a hill, in lee of the wind. A campfire can also be moved if necessary.

It is best to build a fire pit on soil or sand, a good distance from any vegetation that might burn. It is easy to scrape away the sand or soil and make your fire pit in the depression. When you are ready to leave, you should extinguish the fire and cover the area with the original soil or vegetation, leaving no trace behind.

Building a fire on rock

A small campfire will burn nicely on a flat rock. If possible, flip the rock over before making a small fire for your coffee kettle by piling small stones around it. When the fire has died out, you can douse the embers effectively with water and turn the large rock over again, thus leaving no signs of your visit.

Fire on sand and soil

It is easy to dig a shallow pit in an area with sandy soil, such as a pine grove.

If you intend to build a fire on soil or moss, make sure the area near the fire pit is free of roots or ling/heath. Roots grow below the soil and easily smoulder and catch fire, even if you think you've extinguished your fire properly.

Dry ling only needs a spark to catch fire, and some kinds of ling burn at an explosive speed.

Peel away the layer of soil/vegetation where you intend to build your fire. When your fire has died out, simply douse it with water and cover it with sand, soil or the vegetation you peeled away.

Building a fire on swampy soil

A moist bog or marsh is a safe place to build a fire. Peel away a layer of turf, exposing a shallow pit. It will be quite wet, so lay dry sprigs or branches on the wet ground as a foundation for your fire. When your fire has died out, you can simply douse the embers with water and cover the pit with the turf.

Campfire or fire ring?

Many people do not like the idea of leaving a ring of stones behind when visiting natural zones, unless the fire ring will be permanent. A fire ring is an obvious intervention in nature – it will be visible for ever. A ring of stones is actually unnecessary: it blocks much-needed airflow to the firewood. To use a medical expression: it obstructs the airways.

If you intend to build a fire pit next to a permanent lean-to or year-round campsite, the best option is to give the fire pit a stone foundation. We can take a Sami fire pit as our inspiration: clear the ground of ling, grass and soil first. Dig a shallow pit and lay down several flat stones as the foundation. Lay a ring of stones around the foundation and fill the cracks with small rocks/sand/soil.

This reduces any risk of sparks igniting nearby turf or ling, and the stones will also reflect and store the heat, which means the fire will stay warm for longer. The Sami people utilize this kind of fire pit at their camps, in turf huts (*gammer*) and tents (*lavvu*), where it is important to keep fires under control so that the reindeer hides near the pit (*àrran*) do not catch fire.

Igniting a fire

Getting a fire to burn is an art form. The secret lies in getting the first flames to rise. This phase is decisive. Preparation is of the greatest importance. Before the first match is lit, have all your prime materials within reach beside your small piles of firewood.

Use birch bark, juniper bark, dry grass and thin, dry spruce or pine twigs as tinder. Add more birch bark and several thin, dry twigs once your tinder starts burning. Be patient and allow the fire to catch the thin twigs before adding thicker sticks.

Keep an eye out for firewood

I am in the habit of looking for dry firewood as I hike. I look for dead trees with dry twigs and branches. That way I know where I can find suitable firewood before I search for a good spot for my fire. I occasionally gather tinder and kindling as I move along. I also keep my eyes open for big branches and logs so that I do not need to waste time searching for them once I get my fire burning. When I find dry logs and thick branches I usually drag them with me to my campsite so that I can chop and split them near my fire.

She tended her fire, she took
short walks in search of dry wood
and leaves, stocking the fire when
it lagged, squatting like a golem by
its warm perimeter
and meditating on the flames.

Gil Adamson, from The Outlander

One simple rule to remember:

Bark and chips below, dry kindling above, and feed the flames as it catches fire.

Whittle at the sides of your branches to "feather" the surface, which helps them to burn quicker. This also allows the flames to catch and start the burn process.

Keeping it burning

It is a good idea to gather enough spruce and pine kindling first. Thick trunks can be cut into short lengths and split into thinner sticks. I occasionally walk past a suitable old pine tree that I leave untouched because it has given so much beauty to the forest for many decades. There is no reason to fell a large tree just to fuel a small campfire. There are enough dry branches out there and you will be surprised how little fuel you need to keep a fire burning.

Firewood

Norwegian law protects our right to move about the countryside, but that does not mean you can do as you please. The so-called "freedom to roam" is a part of our culture, but there are rules that tell us if and how we can use old trees as firewood. We cannot simply take what we want. Remember to think about how much firewood you require and gather only what you need. Respect the forest and other people's property.

Getting a fire going in the rain

Many would say that there is no point in trying to light a fire in the rain. They are wrong. It can be challenging, but not impossible. Campfires provide radiation heat. In light or normal rain, the heat will not be hindered by raindrops. Sitting beside a campfire in fog or drizzle is like sitting beneath a large umbrella.

I always make a fire when it rains and I never have any trouble finding enough dry wood, even during bad weather.

Search for dry branches on spruce and pine trees. The old trunk of a pine tree can provide excellent fuel. Chop off some woodchips. If the trunk smells of turpentine you can be sure it is full of resin. We call that "fatwood". It ignites easily and burns slowly, even in pouring rain.

Resins from pine and spruce always burn well and the small branches make excellent fuel, even when wet. Scrape and cut wet resin from the wood using a knife and smear it on your wet twigs. The twigs will catch fire as soon as you put them on your initial flames; the heat will dry the wood sufficiently to ignite it and keep it burning.

You can also peel dry bark from a birch tree if the bark appears to be hanging loose. Make a small pile of bark and dry spruce twigs, then use the resin-impregnated woodchips to build a tepee on top of the pile and leave a small opening for the wind. Light the pile from this opening. Shield the fire using your body to control airflow. Keep feeding the fire with dry twigs and branches after it catches – being careful not to suffocate it. Remember that fire needs oxygen to burn.

Things to avoid when starting your fire:

Wet or green wood from live trees

Find dry firewood. Wet wood burns poorly, and releases a lot of smoke and little heat.

Thick wood

Requires ample heat from below to burn. Try to avoid using large logs or trunks. If you are unable to find small branches, split logs into thin pieces.

When firewood is sparse

Make sure you gather enough wood, as one or two big logs are hardly enough. Try to gather enough wood before building your fire, or you will be forced to search for more while your fire dies out.

Not enough air

Make sure you build a fire that allows enough air into the core.

At Fireside

I have never seen you so young and beautiful as now
at fireside, gazing into the embers.

[...]

You smile!
At the dancing flames
you do smile: a butterfly, white as snow, wild
and crazed by happiness and falling into that fiery
rose!

Yet I must collect more wood so the fire yet burns...
I have never seen you so young and beautiful as now.

Rudolf Nilsen, from a collection of poems entitled Everyday (1929)

TOOLS FOR TENDING A FIRE

Many campers can start a fire equipped only with a box of matches and what Mother Nature provides. I prefer to carry extra equipment in my rucksack.

Modern tools can get a fire started in seconds. Some campers prefer a fire-striker to matches. I always pack a watertight box or bag that contains a box of matches or a storm lighter with some birch bark and resin-impregnated woodchips.

Matches

Standard matches work fine, but a campfire is easier to light with the longer, thicker ones because they burn a little longer. You can buy these at most grocery or hardware shops. Long fireplace matches are also handy. You can shorten these to make them easier to pack. Matches can be stored in a watertight plastic box, along with the striking surface torn from the original box.

When I was young, rolls of film were sold in plastic pots with sealed lids. Many campers used these to store matches and striking surfaces. Now you can buy special plastic matchstick containers that have the striking surface on the container itself.

If you need something that burns even longer at a higher temperature, you can buy a box of so-called "fire-lighters", which are match and ignition stick in one. They are easy to light and burn for a few minutes each. They were designed to light ovens and stoves but, of course, they also function well outdoors.

Storm lighter

Many campers are tired of finding wet matches in their pocket, or a worn or useless striking surface – or matches that cannot withstand a breeze. Those campers use a lighter. But lighters can also fall victim to moisture. A storm lighter fuelled by butane or liquid fuel, however, functions well in most circumstances. This technology has been developing rapidly, so a modern storm lighter is a good investment. Don't forget to fill it with fuel before leaving home!

Flint-spark lighters

The flint-spark lighter has become a popular tool for hunters and outdoorsmen. It is easy to see why; it is a lightweight and durable tool but, above all, it is unaffected by moisture. A flint-spark lighter is a simple piece of metal containing magnesium. You create sparks by dragging the lighter against a piece of steel or your knife.

Igniting a fire without matches

Many campers enjoy the challenge of starting a fire without a match. It is actually not difficult, but it is even more impressive once you have mastered the technique. The secret is good preparation and a lot of practice. First, whittle some thin shavings of dry wood and lay some torn pieces of birch bark, grass, dry leaves or clothing lint on the ground. Set your fire-striker against a stone or log and scrape rapidly along the steel so that sparks hit this very fine material. Once you get an ember, add more shavings, bark or lint. Blow very carefully to get a flame.

He stripped the mitten from his right
hand and fetched forth the birch bark.
The exposed fingers were quickly
going numb again.
Next he brought out his bunch
of sulphur matches.
But the tremendous cold had already
driven the life out of his fingers.
In his effort to separate one match
from the others, the whole bunch fell
in the snow.

Jack London, from the novel To Build a Fire

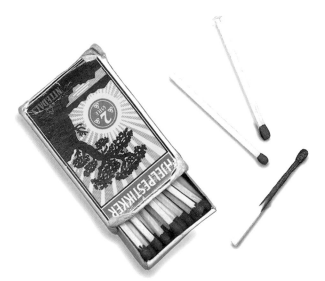

From bow drills to modern lighters

Archaeologists are pretty certain that man was making fire in Scandinavia during the Stone Age (10,000–1800 BC).

Stone Age men probably used some kind of friction method (e.g. a bow drill). This generated enough heat to ignite a dry material, which had to be as dry as bone to catch fire. We later learned that sparks could be made by hitting two types of stone together; some stones created more sparks than others. If the sparks hit a material that ignited easily, a flame could be generated to ignite another flammable material such as grass or dry twigs.

After the invention of metals, it was found that hitting a stone containing quartzite against a bronze dagger or iron auger would make sparks. This was probably how fire-strikers came about; they were widely used in the Iron Age (approximately 500 BC) and up to modern times.

Fire-strikers

The main ingredient of a fire-striker is iron. Iron can get very hot. A striker will release iron shards when it is hit against the sharp edge of an even harder material like flint. Small red-hot shavings are generated when iron is hit or scraped rapidly against such a stone. The trick is to aim the sparks into an easily combustible tinder (amadou fungus was often used). We know that amadou was the most common tinder fungus used with fire-strikers in the Viking Age. The most common of these fungi was hoof fungus (*Fomes fomentarius*), which grows on old birch trees.

Flint-spark lighters

Modern fire-strikers are called flint-spark lighters. They are made from an alloy of iron and magnesium, which generates large quantities of very hot sparks.

As distinct from matches and lighters, flint-spark lighters are not affected by either water or freezing temperatures. Many hunters and outdoorsman consider a flint-spark lighter to be the best fire starter of all, especially in bad weather.

The steel scraper is a modern replacement of flint. By dragging the scraper against the lighter, small bits of steel are released that turn red-hot from friction.

Tinder fungus is traditionally used to catch the sparks, but many campers bring some lint from the clothes dryer, or they use tampons or char cloth (charred linen) instead.

Matches

The objects we call matches were not invented until 1827. The head of a match ignites by means of friction when it is dragged across any rough surface. Chemical matches became immensely popular, but they were dangerous. They could catch fire inside a man's pocket. So-called safety matches appeared on the market in 1844. They contained less phosphorus and would only ignite when dragged across a special surface. In the 1800s these matches began appearing on the market in special boxes with the striking surface on the outside, similar to the matchboxes that are sold today.

Tapers were much used in the Middle Ages.

Nitedal Aid Matches

The first Norwegian matches were produced as early as 1838. Many match factories popped up around the country in those years.

The Nitedal factory was founded in 1863. In 1941 it opened a new factory to make "Aid Matches" (*Hjelpestikkene*), from which a penny or two from each box sold was donated to humanitarian aid.

All match production in Norway moved to Sweden in 1984 after Swedish Match Industries took over production, including that of the Nitedal factory. Swedish Match Industries is still the world's largest manufacturer of matches (having 14 per cent of the global market share). Annual production is impressive, with eight billion matchboxes (390 billion matches) being sold in 145 countries.

Lighters

Lighters generate a spark on a small piece of metal scraped against a tempered steel wheel. The earliest lighters sent a spark into a wick infused with naphtha – this is still found on the market in the form of the legendary Zippo lighters. Most modern lighters use butane as a fuel.

The Sami elders in Udtjá showed great interest in the hearth. It was their place for fellowship and enjoyment. They were not much use anywhere else, other than tending the fire. They gladly tended the source of their stamina and vigour. Fire provides heat and light; they tell us fire is life itself. The elders were happy to watch the fire burn. They wanted the fire burning on the warmest days, even when cooking was not necessary. They would say, "We need our fire to be happy!"

Anders Larsson-Lussi

A sheath knife makes a good whittling tool, but it is a bit weak for splitting wood.

Knives

Many campers would never leave home without a knife. A knife is an essential tool if you need to make shavings, cut twigs or whittle a branch to move a hot kettle from the fire. A simple sheath knife is usually sufficient, but the blades of these knives are often too short for cutting bread or handling fuel. A good knife should have a long thick blade, and the handle should be easy to grip and make your knife feel heavy. A knife like that is a universal tool for hunting and fishing. Or it can be used to gather, whittle and split wood.

The Brusletto Rondane is an excellent camping knife with a full tang blade (i.e. the width of the blade does not shrink inside the handle). This design allows you to hit the blade on the back using a log without damaging the body of the blade inside the handle. A standard sheath knife has a thinner, narrower blade running into the handle. That makes the tang itself a

weak point. If you use a knife to split wood by hitting the spine with a log, you may bend the tang and damage your knife at the heel. This does not happen if you carry a traditional knife such as the one used by Sami people. It can take a beating and be used to cut bread or as a spatula for frying, as well as to whittle, chop and split wood.

A Sami knife often has a birch handle, a brass ferrule, a long rough-forged blade and a hide sheath. The blade is usually 20–23cm (8–9in) long and forged in the traditional manner from carbon steel. Their size and weight make these knives handy and versatile tools. The outer edge of the blade can be used for delicate and detailed work such as slaughtering, filleting, whittling and working with hides. The heel (the part of the blade closest to the shaft) is used to split wood or hit hard objects, while the middle section is used for splitting branches.

A full-tang blade or Sami knife works like an axe in many situations if it is heavy enough to split wood. Saws are also handy on a hike.

Norwegian outdoorsman Lars Monsen has developed a modern knife in co-operation with Brusletto, which is inspired by the traditional Sami knife. It has a heavier shaft and thicker blade, with properties similar to an axe. It can also be used for whittling or gutting fish. The spine of the blade has a groove which can be used to remove a hot kettle from a campfire. The Monsen series, Sami knives and Swedish Fallkniven are all large knives that weigh heavy on the belt, but go unnoticed if carried in a rucksack or backpack.

Axes

You will probably not need an axe on a hike, but a hand axe will come in useful on camping trips and at campsites generally.

A hand axe should have a relatively long shaft and a heavy oblong head. A hand axe is an effective tool for gathering kindling, chopping branches or splitting logs, and, despite its size, it can even be used to fell small trees. Short axes are useful for splitting wood, but that is about all they are good for; they can usually be replaced with a robust hiking knife.

Saws

Even though a good hiking knife will suffice in most situations, some campers like to pack a small hiker's saw as well. A small saw is very effective at cutting green or dry wood, as long as you don't need to cut thick branches or trunks.

A collapsible or folding saw is very handy. It takes up little space in your backpack and is quite effective if you learn to use it correctly. Most saws are more effective when pulled, and remember the timeless tip of letting the saw do the job for you. In practice, that means not using too much force or pressure when sawing.

Your Ignition Kit

If you wish to start a fire under any conditions, always carry these items in your bag:

Matches

Choose extra-long matches (like those used to light fireplaces), which you cut into sizes that are easy to transport.

Strips of Birch Bark

Thin Woodchips Infused with Resin

If you do not carry resin-impregnated woodchips, you can replace them with thin, dry spruce twigs or dry grass.

Fireplace Starters

Fireplace starters are bags that contain a substance similar to paraffin mixed with petroleum and bio-oil. They are very effective, take up little space, are non-toxic and do not stain clothes or gear. A box of homemade starter (sawdust mixed with paraffin) is also a helpful supplement for bad weather. Pack everything in a sealed, watertight box.

FIREWOOD FOR CAMPFIRES

After getting acquainted with the different kinds of wood in the forest, I began to learn how each could be used as fuel for my fire.

Dry pine or spruce both produce a lot of sparks when they burn, but they are definitely my favourite types of firewood. If I want a campfire that does not crackle and pop, I choose deciduous trees instead.

Kindling

You will need wood that burns easily to get a fire started. Twigs or branches from live trees are of no use except for small birch and juniper twigs.

Birch bark burns extremely well, even when moist. So do the basal duff and twigs found at the base of large spruce trees when they are dry. Break up a handful of small, grey twigs and put them in a pile. You now have the perfect kindling ready to catch fire with minimal flame.

Dry twigs are easy to find on spruce and pine trees. Birch trees that have fallen are often dry at the fracture. The wind dried the wood as soon as the tree fell. Even if a trunk is rotten, the splintered areas that were left unprotected by bark will still be hard and dry.

Spruce and pine can be sparse in high mountain areas, and other types of wood are non-existent.

A campfire begins with tinder and kindling. The fire needs fuel as it grows; feed it gradually with slightly bigger pieces.

It is good to know that some types of wood will provide fuel even when they are rotten. Dwarf birch burns well, even when it is green and frozen. So do green birch twigs, especially if you start a fire with an ample supply of tinder, a hot core made of dry woodchips and enough airflow into the core.

"Witch's hair" (lichen, i.e. *Alectoria*) hanging from trees makes good tinder in dry weather. If the humidity is high, the "hair" rots quickly, so do not waste time trying to get it lit.

If you are planning to stay a few nights at the same camp, it is possible to make your own kindling from green birch twigs. Pick some birch twigs and tie them in small bundles. Lay the bundles on a dry spot next to the fire or hang them above warm embers. A bundle of birch twigs can rest directly on the embers as well, but one minute on each side is enough or it may burn. The bundle does not have time to catch fire in that time and will be dry enough to use the next day.

My parents always used newspaper to light the oven. My father also carried newspaper in his rucksack. He used newspaper to start campfires too. I learned some years later that birch bark has the same effect as newspaper. However, it does not get very wet in the rain, as paper does. Birch bark also takes up less space in your rucksack, is lightweight and certainly the best way to get green birch to burn.

A piece of birch bark found soaking wet on the ground will burn well after a short while. It actually burns better than green bark taken right off a live birch.
– I know, I have tried it.

Nils-Henrik Gunnare

The flame he got by touching a match to a small shred of birch bark that he took from his pocket. This burned even more readily than paper. Placing it on the foundation, he fed the young flame with wisps of dry grass and with the tiniest dry twigs.

Jack London, from the novel To Build a Fire

Birch bark

We are talking here about the outer layer of bark (white) on birch trees – living or dead, standing or fallen.

Bark taken from a live birch is still "green". It can be difficult to get this to burn at times and it generates more smoke than dry bark. Bark can be taken from rotten birch trees because the bark does not rot. Bark taken from a dead or rotten tree is usually thicker than bark from live trees, and it does not matter if the bark feels a bit wet on the back. It will still burn well because it is usually only damp on one side.

Experienced outdoorsmen often collect dry birch bark from old trees during winter, and from young trees in the spring when the bark is easy to remove. When dried and stored it will provide good kindling for the rest of the hiking season.

How to gather bark

Search for a couple of birch trees that are growing straight. Slice the bark around the trunk using a sharp knife. Cut a new slit about 20cm (8in) above the first slice. Slice a vertical slit from the upper to lower slice. Use your knife to carefully peel the whole piece of bark away from the tree as it loosens.

Take bark from several trees and dry it underneath something heavy. This will dry the bark as flat sheets.

In birch trees the water rises from the roots in thin capillaries (tubes) that run through the inner brown bark. If you slice through the brown bark you will cut through the capillaries and hinder water from reaching the branches and foliage. Try to avoid doing this. The tree will then continue to transport water from the roots and a new skin will grow where you peeled off the bark.

The bark from a dead, dry birch tree is hard and brittle – similar to crispbread. It slips easily from the trunk.

The bark comes off the live tree in sheets, both thick and thin. Even the thin sheets can be used to get a fire started.

A good knowledge of the various kinds of wood is helpful. It will help you to decide the best type of wood to use for kindling, for light and heat, and for cooking.

Spruce

A spruce tree dries out quickly once it dies. Even a fallen tree dries while on the ground. This happens because spruce has a thin layer of bark. Moisture from the trunk seeps through the bark when the tree is exposed to sun and wind. Dry spruce burns easily and gets hot fast, but the rising embers and sparks can burn clothes, sleeping bags and nearby vegetation. That is why spruce is not the best fuel for night fires. But it does make for excellent kindling. Thin dry twigs catch fire easily but burn up quickly. It is also worth noting that dry spruce logs are easy to split; a sturdy knife is often enough. Small spruce logs function quite well as fuel for cooking. The heat below your frying pan or pot is easy to control if you use small branches; add smaller quantities of fuel often.

The roots of old, uprooted trees are ideal for cooking because they do not produce sparks, as other firewood does.

Decayed spruce trunks also serve as good fuel, once the core of your fire is hot. The dry, decomposing wood burns slowly without much popping or crackling. This is a good source of fuel for night-time fires, even though the smoke can be bothersome.

Pine

Dry pine is a good fuel and never far away in a dense forest. Finding an old dry pine tree is never difficult, and there are always dry branches at the base of a trunk.

Dry pine branches are hard and remain dry in rainy weather. Pine branches can contain resin that provides a good burn, even in heavy rain. Pine often generates large flames. A campfire fuelled by pine sends flames high and makes a good light. As with spruce, the disadvantage of pine is its tendency to throw out sparks and embers. Decayed dry wood from uprooted pine trees will burn well once your fire is hot. Green pine does not burn.

Fatwood

Fatwood is produced when a pine tree tries to repair an injury. Resin rushes to the point of injury when a branch or the bark is broken. The injury gets saturated with resin. Resin-impregnated pine burns especially well and has various names: "fat lighter", "lighter wood", "pine knots" and "heart pine". Chips of resinous pine make for excellent kindling, but fatwood does produce more soot than normal wood. Fatwood is found in the stumps and roots of old pine trees, and occasionally even in the canopies.

A live pine tree with damage to the bark. Pine resin is excellent for kindling – especially when used in rainy weather: simply rub some resin on your twigs before lighting the tinder. ›

Fallen birch makes poor firewood. Wet birch rots quickly and stays wet.

The spot where the birch broke before it fell is often dry. ›

Birch

Outdoorsmen consider birch to be the best firewood of all, whether indoors in fireplaces or outdoors. Having said that, birch is not the easiest wood to find. Dead birch rots as soon as it gets wet, and then refuses to burn. This is because the bark traps moisture inside the trunk, even on a warm summer day. Even if the tree is dead and the wood is rotten, the smaller branches near the crown may still be dry in places where large branches or parts of the trunk have lost their bark. Break off the dry twigs and branches by hand if they are not covered with a thick layer of bark. If they crack when you break them, then you can be certain they will burn just fine. A birch trunk lying on the ground will be moist and rotten below the bark; this is never a good choice of firewood. But the white bark can still be quite useful as kindling.

Other deciduous trees

Dry rowan is a good source of kindling and fuel for your cooking fire. Dry willow is also an excellent wood for cooking. It burns without smoke and smells nice. The disadvantages? Willow burns fast and produces few embers, compared to birch and other dense woods. Beech and oak, which are known for their high BTU output, make poor firewood. First of all, you can almost never find dry oak, beech, maple or elm in the forest, and the green branches of these deciduous trees are not good for firewood. But there are some exceptions – for example, the green wood from ash trees burns well.

Twigs and branches from live ash trees can be used as fuel.

^ The thin twigs at the bottom of
   a small spruce tree provide excellent
   kindling.

‹ Aspen burns well when dry, but it
   is far too moist when green.

^« Rowan (on the left) and willow
   (on the right) burn well when dry,
   but they are not suitable as firewood
   when green.

« Dry pine twigs are a good source
   of firewood.

Dry juniper twigs are also great fuel. The dry bark from juniper combined with other tinder is great for getting a fire started. Green juniper needles make good fuel but produce a lot of smoke.

Bog myrtle and osier burn well, especially the thin, dry twigs.

Using green wood

Green wood (i.e. the wood from live trees) is useless as firewood – deciduous or conifer. The only exceptions are juniper, birch and ash. Juniper burns quickly with an intense heat, but it produces lots of smoke. Raw ash also burns well, but it is not a common sight in Norwegian forests. In our mountains you will find downy birch and dwarf birch, exclusively, which burn fine when green. So do basket willow (osier) and bog myrtle. This is valuable information, because drier kinds of spruce can often be hard to find.

Downy birch

Adult downy birch grows near the timberline and along streambeds. Downy birch can be gathered without using an axe. The smallest branches and twigs are easy to take from the trees and the larger branches can be removed with a heavy knife. It takes more green wood to maintain a fire than dry wood. You should remember to collect a little more than you think is necessary, as green wood burns faster than dry wood.

Downy birch burns surprisingly well and no special tricks are needed to get it to burn. You should be able to get a little fire started using only matches, bark and a bundle of thin birch twigs.

A birch fire produces more hot embers than dry spruce and pine. This is good to remember if you plan to use your campfire for cooking and frying.

Dwarf birch

Green dwarf birch burns well too, but starting a fire with it requires patience. You need to feed the initial flames one twig at a time, and make sure that the core gets lots of air. You will need a big pile of small branches to keep a small fire burning. A good tip is to use your jacket or the cover for your sleeping bag to gather lots of sticks before starting your fire.

Bog myrtle and osier

Where mountains are bare, you may need to search long and hard among the birch to find bog myrtle and osier. Fortunately, you do not need much to keep a cup of coffee warm or to boil water for your bag of instant food. If you find bog myrtle or osier, you will certainly have enough fuel for a small fire. Small groves or thickets of mixed vegetation will often hide osier and bog myrtle, often near streambeds or small lakes, in glens and dales, or on hills facing the sun in lee of the prevalent winds. These areas often have shrubs with thick stems. The twigs from osier and bog myrtle are easy to break using your hands, but gathering thicker branches and stems requires a good hiking knife.

Bog myrtle (*Myrica gale*) is almost never wet to the core, even if found in moist soil. This means that you will be able to find dry osier or bog myrtle following days of heavy rain.

Juniper

The heartwood of juniper bushes is often dry after days of rain, which makes it a very good alternative to bog myrtle and dwarf birch. Juniper bushes grow close to the ground in mountainous areas, but make an excellent fire. The needles burn with intense flames. We do not worry much about forest fires high in the mountains. Let your fire smoke and crackle if there are few trees. The most important thing is to get a fire going.

Crowberry and ling

When firewood is scarce you can turn to other sources of fuel. Ling and heath always provide dry stalks that can be used as fuel. The dry stalks of crowberry are only a few centimetres long and a few millimetres thick. But, my oh my, how well they burn! If you are lucky enough to find a patch of dry crowberry with dry needle-like leaves, you will have found an explosive source of fuel. Live crowberry burns quite well once you get your fire started, but you will always find dry ling among the live plants. Ling grows slowly and it takes a long time to spread. We prefer to leave it alone to grow in peace.

Crowberry

**Not even in winter,
When we slept on cots in tents while
camping in the forest,
did we prefer dry wood over green birch.**

Isak Parfa

———•••••••◆ ◇•••••••———

Why does green birch burn?
When green birch gets hot you will see the sap boil
and bubble, hissing from the open slits, but the birch
still burns. This is probably because birch sap
contains essential oils.

Green spruce does burn if your fire is hot

Fresh spruce burns too, but not very well. Using birch bark and spruce twigs to get a spruce fire going does not work, but green spruce will burn if thrown on a hot fire made of other wood. It is slow getting started, but the crackling will intensify as the spruce burns.

Green pine is no good

As I said earlier, green pine doesn't burn. Live or green pine is just about the worst firewood of all. A green branch of pine is like asbestos. It will scorch on the outside, but refuse to burn – even when thrown on a powerful fire. This fact makes green pine branches an excellent foundation on which to build a winter campfire; pine logs keep the fire from sinking into snow.

The core is essential

Fire and flame produce powerful radiant heat. The glowing embers of a campfire produce a special kind of heat as the fire dies. The embers release heat in a different way to the flames. Dry pine and spruce usually only leave a pile of ashes. The pile of embers from green wood will stay hot and can be quite big. Even though the temperature is high, throwing more green wood on the embers will not produce more flames. The wood will just whistle and smoke. The wood will eventually quench the embers. If you lay several logs on the fire in a way that allows air to move between the green wood and the embers, it will not take long before flames appear below and on the sides of the logs.

The Sami keep a few large rocks beside the fire in their turf huts and tents. The logs rest against the rocks to create an airspace between the wood and the glowing embers. This creates enough heat to release the gases inside the wood, which burn until the wood itself ignites into flames.

Burning green wood

Making a campfire from green wood requires a little more effort. The Sami people have a great deal of experience in using green wood for their fires. The green logs are laid facing in the same direction.

When wood is laid parallel it burns with less smoke, but you need to feed the fire often with small, dry sticks to keep the heat intense enough to burn. Put a large log on one side to lift the wood above the ground, to provide oxygen from below and along the whole length of the wood. Splitting green logs is a waste of time because they burn just as poorly split as whole; just remember to keep an airspace between the logs and the embers.

Building your fire using small sticks in a criss-cross pattern (a crow's nest) creates too much air between the sticks. This will kill the flames and your fire will die out. Three pieces of firewood lying side by side will burn long and well, while three pieces of firewood laying criss-cross will only burn where they cross. You would need a lot more wood to keep a crow's nest fire burning hot.

Green wood tends to burn without sparks, which makes it ideal for fires inside large tents such as the Sami *lavvu*. The flames stay low and the fire burns calmly with little smoke (except when you are trying to get it burning with leafy birch twigs). We can learn much from the Sami about how to make a fire pit inside a tent or apply this knowledge to modern camping trips where large tents are used.

Tyrielden

(extract from a poem)

She has broken many a pine branch over the years,
With its crown stretching towards the blue heavens,
– the splendour of winter snow and the golden sun of spring
before she abruptly bursts through the winter storm.

The summer sun splitting the trunk,
Until the resin bled, thick and well
To warm the logger's bones
In the cold winter day.

Gratitude, heart pine
for your warm red flames,
and for the summer light you spread
upon fields of frost and snow.

Hans Børli, Kongsvinger Arbeiderblad, 1 February 1942

TYPES OF CAMPFIRE

*A campfire is a gathering point
that provides light, heat,
food and enjoyment.*

> White man makes big fire, stands far away.
> Indian makes small fire, sits close.
>
> *First Nation People saying*

This chapter explains how to build the most popular campfires and what these are most suitable for. The natural surroundings will often determine what kind of campfire is possible. The abundance of dry wood in the area will influence your choice, too. In the high mountains (where firewood is scarce), you must plan to use less wood and it is also wise to shelter your fire from wind and weather. A campsite in the woods that is to serve as base camp for several nights requires a different type of fire than that needed for a rest and a hot cup of coffee.

WHILE HIKING, we usually build a simple fire with the firewood we find nearby. Many hikers make the mistake of building a crow's nest fire, piling up wood in a random criss-cross pattern. As I've already said, this creates too many air pockets around the fuel, which means that your fire will die out quickly. However, if you lay firewood in parallel rows, it will burn better.

Break small spruce branches by hand into pieces 20–30cm (8–12in) long. Lay a foundation on flat ground using four small thick branches placed side by side. Lay a large log to one side of the foundation and rest the ends of the first layer of logs on top. This allows oxygen to get below the first layer and keeps the wood off the embers.

Lay the second layer on top of, and at 90 degrees to, the first layer, again with the logs placed side by side. Then lay one or two sheets of birch bark and spruce kindling on top of the second layer of logs. This campfire is lit from the top. When it catches fire, keep feeding it with twigs and small branches. Once your fire is burning, lay one log on each side of the fire crosswise to the previous layer. This allows you to add a new layer of logs crosswise on top of your fire without smothering it. This technique burns fuel from below and above. Continue to feed your fire by adding logs, laid side by side. This allows good airflow, which will keep the fire burning hot and long.

1) Lay four to five logs, side by side, with one end resting on a log to provide airflow. Build your fire by placing a second layer of branches crosswise on top of the first layer.

2) Lay the kindling between two thick branches placed crosswise and on top of the logs/branches below.

3) Light the fire from the top and let it burn down through the layers.

A TEPEE CAMPFIRE is probably the most common type of fire. Shaped like a tent, a tepee fire provides a lot of heat and ample light.

Build your tepee in a circular pattern, using thin, dry sticks and small pieces of split logs with the tops touching. Make a small pile of tinder and kindling from woodchips, birch bark or dry ling/grass that will ignite easily. Wait until your tepee is burning well before adding larger logs to make it bigger and hotter.

A tepee campfire is suitable for large fires at campsites or rest stops along the trail. It is also good for heating coffee kettles or roasting hot dogs and other food on a stick. I recommend taking the time to build a tripod above the fire by lashing three long branches together. You can then hang your kettle or pot from the tripod (see pages 74 and 98).

A tepee campfire is made of sticks and split wood, in the shape of pyramid, and lit from below, to give a good amount of heat.

A TWO-STONE COOKING FIRE is a small fire bordered by two relatively flat stones of equal size. It is designed to hold a frying pan (for cooking meat or fish) or a pot in which to boil water or heat soup.

The distance between the stones should allow you to rest a frying pan or cooking pot on them, with enough space to make a small fire in between.

Keep your two-stone cooking fire small.

A three-stone cooking fire is good for mountain trips because
the stones protect the fire from wind.

A THREE-STONE COOKING FIRE is a small campfire surrounded by three stones. This campfire does not require much fuel, which makes it popular among campers on short day trips.

As the name implies, the fire is built by placing three stones in a V-shape around an area large enough for a small fire. The three stones control the heat and focus the flames upwards. This type of fire is excellent for simple cooking because the stones act as a stove to hold a kettle or cooking pot.

The three-stone fire does not need much firewood if you only need to boil some water for a cup of coffee, heat a bowl of soup, or warm the trail lunch you made at home. The only fuel you will need is a little bark, some twigs and a few short spruce or pine branches. Your fire is ready for the kettle or pot when the flames are burning evenly with little noise, and a hot bed of embers has begun to settle. Feed the fire with two or three logs at a time, as necessary, to keep it burning.

A KEYHOLE CAMPFIRE is based on the same principles as the three-stone cooking fire, but it has higher walls that act as a wind screen. The walls of the keyhole campfire usually form a U- or V-shape, with the opening on the leeside. The fire itself is a small tepee on the leeside of the stones. Start small and feed the fire with larger branches as needed.

If you intend to prepare meals on a keyhole fire, it is wise to build level walls on which to balance your pot or kettle. In heavy winds you might need higher walls to provide additional protection around your cooking area. You will want to protect your pots from cold winds while cooking.

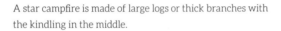

A star campfire is made of large logs or thick branches with the kindling in the middle.

A STAR CAMPFIRE functions quite well for cooking if you cannot find stones or branches long enough to build a tripod (see page 98). Build your fire using a few long logs that serve as both fuel and as a support for a frying pan or cooking pot.

Lay five or six logs in the shape of a star. Build a pile of tinder and kindling in the centre of the star using dry twigs and birch bark. Shave the ends of the logs with an axe or knife (feathered) to help them burn faster. Once your star is burning bright, you can put your pot or pan on the logs. Continually slide the logs towards the centre of the fire as they burn, or you will lose the foundation for your pan.

A star campfire is good for cooking if you have a tripod.

Build your fire by laying logs side by side so that your kindling burns the logs from the top down. The burning bridge drops through the valley walls onto the river below.

A RIVER-AND-BRIDGE CAMPFIRE is easy to get started and burns quite well. The special design of a river-and-bridge fire provides ample airflow below and hot flames above.

Start by building a base of thick logs on the ground, side by side (this is the river). Then place two thick logs on top of the river so that they are facing in the same direction and at an ample distance from each other (these are the valley walls). Pile your tinder/kindling on the river between the valley walls. Then build the "bridge" so that it is resting on the valley walls above

the kindling. You can then cover the bridge with a row of dry branches.

The "bridge" will burn first, thereby igniting the logs on your bridge. The burning bridge will collapse into the river valley and slowly begin to burn the "river" itself. This technique provides ample airflow to the kindling from below and excellent heat and flames from above. The campfire generates good combustion and little smoke. The extra heat helps to burn the thick logs below and on the sides.

A LOG CABIN CAMPFIRE is the best for campsites and cooking. The square shape looks like a log cabin with parallel walls, with each layer of logs/branches at 90 degrees to the layer below.

This type of fire is easy to build. Begin by laying two parallel logs or thick branches on the ground at a distance of 20–30cm (8–12in). Lay two new parallel logs on top of these, at an angle of 90 degrees, to create an empty space inside the log cabin. Lay some tinder, birch bark and other kindling in the centre of the cabin. Use woodchips as tinder around the bark and build a small tepee using the kindling. Once the kindling catches fire, keep feeding it with dry twigs and woodchips before building the rest of the walls around your cabin with a couple more parallel layers of logs. The framing of the walls allows oxygen to enter the core of the fire. Keep feeding the fire with twigs and larger logs.

The log cabin campfire burns hot and consumes a lot of fuel. However, it is a good source of heat and light in both autumn and winter. It also burns slowly, making it a good choice if your firewood is moist or wet. Use moist wood on the bottom layers so it dries out before the fire burns down that far.

The log cabin is great for cooking and can heat big pots of water or soup rather quickly. You can rest your kettle on the walls of the cabin or hang it over the flames from a tripod.

This fire is built on a foundation of two solid logs, layer by layer, with the logs set at 90 degrees to the logs/branches below.

LEAN-TO CAMPFIRES function well if you have a really big log available. This type of fire will burn slowly and is a good choice at night, while you are sleeping.

Build your lean-to fire by laying some branches against a large log or tree trunk. Start a little fire of thin twigs next to the log. Once you get your kindling burning, simply add more pieces of wood by leaning them over the flames, with the ends against the big log in a fan-shape, as shown here. Once the big log starts burning you can choose to build a second lean-to on the other side of the log following the same process.

A lean-to fire is built against a large log or tree trunk.

A TRENCH CAMPFIRE, which is also called a hunter's campfire, is simply a fire between two heavy logs. The fire runs the whole length of the logs and makes an excellent stove and foot-warmer.

Place two long logs, side by side, and then build your fire in the narrow trench between them, using twigs and small branches that rest against the logs. The trench provides protection from the wind so that your match will not blow out as you light the fatwood chips or bark below the twigs. Lay thin dry twigs against one another inside the trench to get the logs burning. The wood burns quickly, so keep enough firewood to hand to feed the fire as it burns. Once the walls of the trench begin to burn, you will need much less fuel to keep the fire burning.

The logs provide an excellent support for cooking, which is why hunters prefer this type of fire for frying meat or fish in a frying pan. If you only need a small fire, use smaller logs.

A hunter's fire is made from two trunks or big logs.

V-SHAPED CAMPFIRES or **PLOUGH CAMPFIRES** are made of two large, round logs or trunks. The plough fire is often used by small groups of hikers to prepare basic meals over a fire.

Place two logs in a V-shape with the open end facing the wind. If you need a large fire, you should use large logs, but the "plough" can also be built with small logs or branches. Regardless, lay your logs/branches in a V-shape. Build your fire at the tip of the V (a small tepee fire is good). Once the fire is burning, feed it from the opening with small sticks. You can build your "cooking grill" by placing two or three thick green branches on top of the two solid, V-shaped logs. These branches will function as the base for your pan or cooking pot, so do not let them burn!

Sit at the opening of the V to stay out of the smoke. You will also feel more radiant heat from this side. You can adjust airflow and the intensity of the heat by changing the size of the opening. This fire will burn longer if you use thick logs (of spruce or pine). Bigger logs reflect more heat towards those sitting fireside or towards the sleeping area.

A plough campfire is made from two trunks or large logs that form a V-shape with the kindling and fuel between the logs.

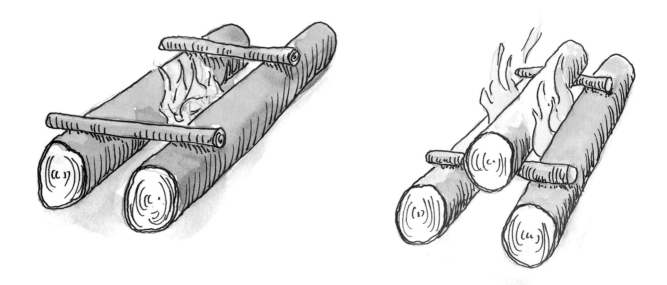

A Finnish gap fire is made from two large logs with a fire in between that extends the length of the logs. Once the fire has caught, simply place a third log above it, supported by two branches across the gap at each end.

A FINNISH GAP FIRE (*rakovalkea*) is the traditional lumberjack's fire. It requires long thick dry logs of pine or spruce. This fire burns very slowly and stays hot. It is an excellent night-time fire for campers sleeping inside a lean-to or under the stars.

Use long dry pine or spruce trunks/logs. Start by placing two long logs on the ground. The space between them should be equal to one log. Build a small fire between the logs using easily combustible branches to get a good fire burning. Once the fire is

hot and the logs begin to burn, lay two heavy branches crosswise to span the gap at both ends of the logs.

Place another long log on top of the two branches. It is a good idea to cut slits in the logs along their length on the inside of the gap. This makes it easier for the flames to eat into the logs. The longer the logs are, the longer and hotter they will burn. The lumberjack's campfire does not need to be fed constantly, which makes it a great fire to have while you are sleeping.

THE SWEDISH FIRE TORCH or **CANADIAN CANDLE** is nothing more than a thick log with deep cuts at one end. The cuts are filled with tinder and kindling. When lit, these begin to ignite the log, which then burns downwards slowly. This fire provides heat, light and a flat cooling surface due to its solid base.

A Swedish fire torch is made from a thick log, about 40cm (16in) long and 20cm (8in) wide, or maybe a little wider. Saw deep cuts at one end of the log and fill them with some kindling dipped in paraffin, naphtha or lighter fluid to get the log burning. The fire burns downwards, slowly eating into the log – this provides a constant cooking temperature. The torch makes an excellent stovetop because the wide base on the ground or snow remains cold. The flat surface is great for supporting a frying pan or cooking pot.

Finnish soldiers used these logs as stoves during the Winter War. They dipped rags in paraffin before stuffing them into the cracks to help get the so-called "Finnish stove" burning.

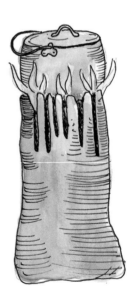

You can make a single big log into a fire and stovetop by cutting deep slits at one end.
Once the cuts catch fire, the wood will burn downwards into the dry log.

A platform fire combines the hiking fire with the log cabin campfire. Build a solid foundation wall or "raft" of green wood and your winter fire will burn hot without sinking into snow.

A PLATFORM CAMPFIRE is considered by many to be the best of all campfires in winter. It is also known as **A WINTER CAMPFIRE** because it is built on a solid log/branch platform that keeps the fire from sinking into melting snow.

If you wish, you can clear away the snow first and build your campfire on bare ground or, alternatively, you can build the platform on the snow (if the snow is hard enough). Start building the platform by making a "raft" of logs on the ground. Lay a row of logs, side by side, without any gaps between them. Lay the second layer of logs on top of the platform at 90 degrees to the first layer. Some campers prefer to use green or wet logs for the second layer. This allows the fire to burn downwards more slowly, which means it takes longer for the raft to sink into the snow.

Unlike most of the other campfires, a platform fire is ignited from the top. The easiest way to get the fire lit is to build a small log-cabin fire on top. Once the log cabin gets burning, it will burn slowly into the layer of firewood below. A platform fire almost never needs feeding.

If you are a little careful and you keep an eye on it, you can even set a cooking pot on the burning coals. Be aware that your stove will become more unstable as the logs burn.

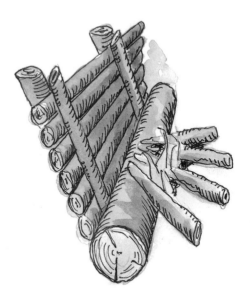

A lean-to campfire with a reflector wall

A REFLECTOR CAMPFIRE provides radiant heat for those sitting or sleeping near the fire. You can use a reflector wall with most types of campfire for extra heat at permanent campsites.

It is an effective way to utilize the radiant heat from a fire. The idea is to use a big rock or wall of logs to reflect the heat from the fire towards people nearby. Keep in mind that the reflector logs can catch fire. It is therefore best to use green or unseasoned wood, which will not burn even if the flames dance along the log wall.

The fire itself can be built in many different ways. The reflector is not a fire itself; it is a screen that reflects the heat. The reflected heat can be so intense that it can even be used to roast fish.

A SEVEN-STONE OVEN is a construction made from seven or more stones and designed as a combustion chamber with an oven above it – not unlike a baker's oven. The purpose of the oven is to bake or roast meat and fish.

You can use more than seven stones, if you wish, but seldom fewer than that. The point is to build a baker's oven. It is a good idea to use large flat stones. This is a two-storey construction with the fire occupying the first storey and the oven the second storey.

Find several flat stones to build a two-storey chamber, with walls on the sides of both storeys and a wide flat stone to divide them, which serves as the floor of the second storey. A thinner floor will get hot faster, and you can even fry food directly on the floor. If you place two stones on each side and use one flat stone as a roof over the oven, the heat from the roof and walls will reflect heat back to your food. The roof will also serve as a warm surface to keep coffee/food warm – you can even light a small fire on the roof if you like. This sends heat from above and below, just like a real baker's oven

A seven-stone oven is built using flat stones:

1) Begin with two stones and a flat stone. Add two new stones to the top surface of the flat stone.

2) Keep building, using one more flat stone. Use another flat stone to make the back wall.

3) Light your fire in the lower chamber.

THE FIRE SITE

Humans have always been drawn to fire. Fire has a magical appeal. The fireside has been a natural gathering point for millennia. We needed fire to make our food, and we gathered around it to stay warm and listen to tales among our people. The hearth is no longer the centre of the home, but we are still drawn to its light and flames, even when it burns behind the glass front of a modern fireplace.

The campfire

The campfire has always been the most basic source of fire, ever since humans lived in small nomadic groups more than 10,000 years ago.

Archaeologists refer to the earliest fire sites as fire pits, which were little more than traces of charred earth. They also found slightly more complex pits surrounded by stones. A circle of stones around a fire protects the surrounding area from flying embers and runaway flames. The stones also absorb heat and stay warm after a fire has died out.

The truth is that neither of these fires was very effective as a source of heat; a campfire only heats the side of the body facing it. Still, the campfire has always been an asset to nomads. A fire lay is quick to build and provides much-needed light and heat once the fire gets hot.

A warm gathering place

Humans are naturally drawn to fire for practical reasons and to fulfil emotional and spiritual needs. The fire kept us warm. It allowed us to make offerings to our gods, burn our dead and punish people found guilty of misdeeds.

In turf huts and tents, it is best to use ›
firewood that does not produce sparks.
Do not use spruce or pine. Green or
unseasoned birch is best.

The fireside was a place for meals, tales and song. Knowledge and experience passed from generation to generation in the light of the flame. We no longer use a fire to fulfil our spiritual and practical needs in our part of the world, but we still seek its magic in the company of others, whether it burns as a campfire, bonfire, hearth or modern fireplace behind steel and glass.

Open fires in tents, turf huts and houses

Humans lived as hunters and gatherers during the Old Stone Age (10,000–4000 BC). That lifestyle kept us on the move. The climate was much milder than it is today. A tent made of hides or a simple turf hut would provide ample protection. Nomads often returned to the same locations. Their camps and tents were designed for rapid assembly; they needed to get a roof over their heads and a fire started.

Reindeer hunters from the Old Stone Age left traces at Sumtangen, a little southeast of the Hardangerjøkulen Glacier, of what is believed to be Norway's oldest dwelling. Carbon dating helped archaeologists and historians determine that this dwelling was in use 8,000 years ago. The walls were probably clad in turf and grass, and perhaps even hides. The floor of the dwelling was round with a fire pit in the middle.

Traces of a similar turf hut (*gamme*) or tent were found on a ridge near the sea at Gamvik, Finnmark, and there are several similar settlements in Troms and Finnmark with fire pits surrounded by a ring of stone.

87

Reconstructed longhouse from Borg, Lofoten

Sami elders tell us these traditions have been kept alive through generations, and they are still in use. They identified their settlements by the circle of stones when they returned year after year. Many of these primitive settlements still had tent poles standing around the fire pit. This made erecting a tent, laying a fire and rolling out some hides easy.

These settlements were more or less permanent during the Bronze and Iron Ages (1500 BC–AD 1030). Archaeologists call the big farmhouses from these times longhouses. The name is befitting; these large homes could be 50m (164ft) long. The largest longhouse ever found in the Nordic region was in Vestvågøy County, Lofoten. It was 83m (272ft) long and 8m (26ft) wide, and had many rooms. It had a shed for the animals at one end and various storerooms, as well as a large living space surrounding the main room, which served as the chieftain's hall. The fire pit was the heart of the main hall and provided light and warmth for those inside. The fire pit was also the kitchen. There was a hole in the roof above the pit for the smoke to exit the building. The fire pit was where the villagers would gather. This is where most indoor activities took place.

The fire pit was always at the centre of the longhouse, from the Bronze Age to the Viking Age. Although the design of the fire pits varied, they all had features in common. They were made of stone to maximize firewood consumption and provide heat after the fire had died out. If it were possible to shut the smoke hole, the heat would remain in the main room for longer. Many were simply made of flat stones laid on the floor, but we also find traces of more advanced kitchens with combustion chambers – not unlike what we would call a baking oven.

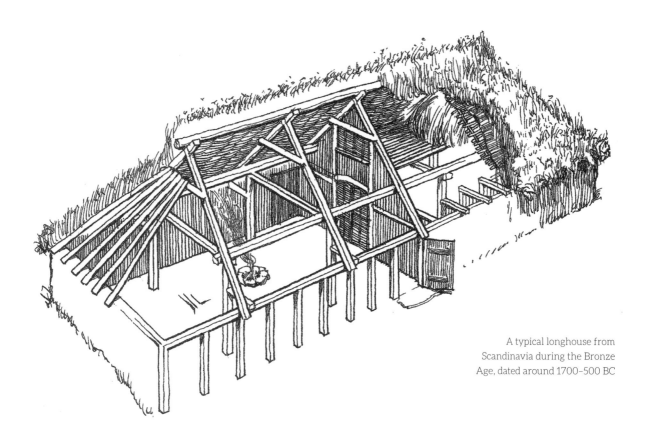

A typical longhouse from Scandinavia during the Bronze Age, dated around 1700–500 BC

The floor-hearth had no chimney. This was the prevalent design in Nordic countries until the end of the Middle Ages (until about AD 1550), when we begin to see stone hearths being raised off the ground. The so-called smoke kitchen used less firewood because it only needed to be lit a couple of times a day; the stone retained heat for hours. Smoke kitchens were usually built in the corner of the room. Clay was used as mortar. A wide opening faced the room and there was still no chimney, so the smoke left the room through the smoke hole in the roof. Although the fire was only lit a couple of times a day, it was loaded with plenty of firewood each time. The cook could use the hot coals to prepare the day's meal once the fire had died down.

The hearth stored lots of heat and kept the room warm all day. The open hearth was replaced in the 1800s by fireplaces with chimneys, or a cast-iron stove with a chimney pipe.

Hearths, fireplaces and stoves

Hearths are generally used in kitchens for cooking, while fireplaces are found in other rooms.

Smoke bays, chimney pipes, chimney stacks, tiled stoves – all of these allowed us to master fire and heat, and get smoke out of the room. Chimneys and pipes made it possible to construct houses with several storeys.

The Scandinavian Hearth

The simplest kind of hearth is the fire pit, which is square in shape and consists of flat stones on the ground in the middle of the room. If the combustion chamber is enclosed in stone and raised above the floor, the fire pit becomes a hearth (since Roman times). Most fire pits did not contain stone, so the pit could be cleaned of its ashes by digging out the soil. Smoke from ancient fire pits and hearths exited through a hole in the roof above.

In a cottage like this, all living things and fresh foods would stay close to the fire, for light and warmth. Benches would be kept close, along the long walls, with the seat of honour in the middle: a guest's worth could be judged by his proximity to the hearth; the best places to sleep were nearest the fire; *Vesta*, goddess of the home and hearth, rules everything; even the dead are carried three times round the hearth before being laid to rest.

Troels Lund, from the book Daily Life in Scandinavia in the 16th Century

‹　Painting entitled *Taking Leave* by Adolf Tidemand, reproduced as a lithograph in the 1858 edition of *Norsk Folkelivsbilleder*. The picture shows a typical Norwegian cottage; there is no chimney over the hearth. The smoke escapes through a hole in the roof.

COOKING OVER OPEN FLAME

Humans discovered the benefits of fire for cooking at a very early stage in our development. The first true kitchen was a campfire.

Most people associate cooking over an open fire with camping, but the joys of food and fire can also be had at home in the garden, at a summer cabin by the beach or at a winter cottage in the forest.

Campfire as outdoor kitchen

A charcoal barbecue grill is not designed for the heat generated by firewood. An outdoor fireplace, however, has space for lots of firewood that generates plenty of hot coals. This makes outdoor fireplaces ideal for cooking and having fun.

You can build your fireplace with brick or stone, or purchase an outdoor kitchen designed for firewood and cooking. You should check whether you need a permit to build a fire pit or outdoor kitchen, and you must speak to your neighbours before starting; smoke and odour can be disturbing if your kitchen is in the wrong spot.

A garden fire pit requires much more space than an outdoor kitchen. Your pit must be 4m (13ft) or more from a neighbour's property and far from other buildings or vegetation. The fire pit needs a good foundation of stone, sand or gravel. It should be enclosed with stone or brick and be wide enough for a large fire.

A fire pan is a good alternative to an outdoor kitchen or fire pit. Fire pans are available from specialist shops in many shapes and sizes. Some even come with grilling grids and fire screens. They are portable and can be set up anywhere in your garden or near your cabin. You can bring them on trips or hikes, as long as the equipment is not too heavy and you do not have far to walk.

You should treat a fire pan as any campfire, even though it keeps the fire and firewood off the ground.

^ Fire pan

‹ Garden fire pit, lined with stone

94

Over flames or embers?

Food can be cooked in the heat of flames or from the radiant heat of the embers, or laid directly on hot coals. Open flames can burn food on the outside, while leaving it raw on the inside. The radiant heat from flames or embers allows the heat to penetrate the food both inside and outside. This produces the best results.

The easiest approach to fireside cooking is to skewer your hot dog or meat in the radiant heat of the flames. However, the best method is to place a grilling grid just above the hot embers. You can also try packing your food in tinfoil and laying it right on the embers.

Food on a stick and food in a pot

Archaeologists in southern England have dug up things like stone axes, animal bone and wood-charcoal. Science has determined that man has been heating food over fire for a very long time, perhaps as far back as 300,000 BC, long before *Homo sapiens* walked the earth. It would appear that they roasted on sticks over open fires or coals. The art of cooking, however, did not arise until much later. Man had to invent pots to hold the food first. Historians believe the oldest clay pots used for food preparation are 13,000 years old. They were found in Japan and were used to boil water and heat food.

Cooking requires putting the food in a container such as a frying pan, pot or kettle first – and then finding a way to hold the pan or pot over the fire.

Two forked sticks and a crossbar

Find a couple of branches from a live deciduous tree that have a Y-shaped fork and push them deep into the ground, one on each side of your fire pit. Rest a sturdy green stick on both forks to form a crossbar. The crossbar should be strong enough to support a small pot.

Spit kitchen

You may need to consider other options if you are camped on rocky ground and there is no place to plant your sticks in the soil. One option is to raise a solid stick above the fire, held in place with stones. Rest the middle of the stick (the spit) on some stones to lift your pot the right distance above the flames. Use heavy stones at the base to hold the spit in place. Your spit should be strong enough to carry the weight of the pot. You will not need a big fire to get water to boil if it is hanging just above the flames.

Two forked sticks and a crossbar

Spit kitchen, supported by stones

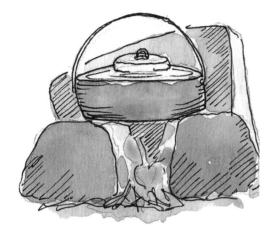

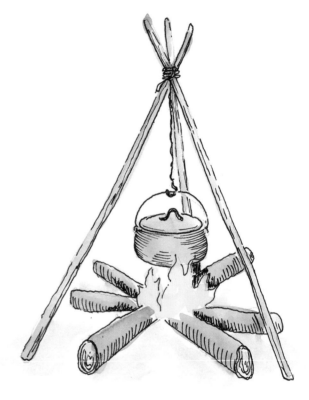

Stone stoves

Set two large stones of equal size on both sides of the fire and rest your pot on the stones above the flames. Use a third stone behind them as a back wall. The third stone should be flat and a little larger than the other stones, since it will function as a back wall and also as a wind screen. Lay the fire in a way that allows the flames to touch the pot. Contact between flame and pot will heat your food much faster.

Earth ovens

If branches and stones are hard to find, you can simply dig a hole in the ground, fill/surround it with dry firewood and light up. It will not take long for the water to boil if the pot is buried in hot coals.

Cooking tripods

Tie three straight, solid branches of birch, willow, aspen or rowan together at one end and set them up like a tepee. Tie them together tightly so that your tripod is sturdy. A tripod is the perfect way to hang a pot or grill over a campfire. Rope can burn, so pack a couple of metres of steel wire (or a chain with a hook at one end) from which to hang your pot over the fire.

Pit ovens

Pit ovens have been around since the Stone Age. A pit oven is a hole in the ground, in which food is cooked using rocks heated by fire. The method is simple and makes meat delicious and tender. It requires some time, though.

Start by digging a pit in the ground. Cover the walls of the pit with stones and light a big fire inside. Let the fire burn out, leaving only red-hot coals. The stones become so hot that they can be used to bake food for several hours. The bigger the stones and the hotter your fire, the longer the stones will retain their heat.

Most campers prefer to remove the hot embers before putting the food in the pit. You may need to protect your food from soot and charcoal remains. You can do this by wrapping your food in baking or greaseproof paper, then in some tinfoil.

Next, surround and cover your food with the hot stones before overlaying the pit with moist turf, twigs, hay or dirt. The heat from the stones will roast your food, and the soil surrounding your oven will act as insulation. A pit oven is hot enough to roast a large fish or cut of meat over several hours.

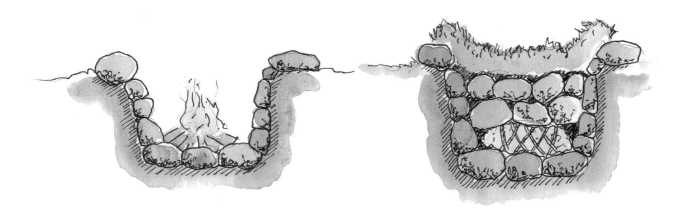

Pit oven

Pit ovens have been around since the Stone Age (10,000 BC) to make meals, extract oil from fish, boil blubber from whales or seals (stone-lined pits) and cure hides, etc. As time went on, pit ovens were also used to bake clay in the process of making ceramics.

Smoking fish and meat

Smoke has been a means of preserving food since the Stone Age. You can smoke meat or fish by hanging it in an enclosed space filled with smoke until it is preserved. There are differences between cold, hot and warm smoking. Cold smoking requires a smoke temperature of 27°C (81°F), but no more than that. Hot smoking requires a smoke temperature of 65°C (149°F) plus, which cooks the meat or fish in the process. Warm smoking is an intermediate process, preserving meat or fish at a temperature of approximately 40–50°C (104–122°F).

‹ Cold-smoked Arctic char

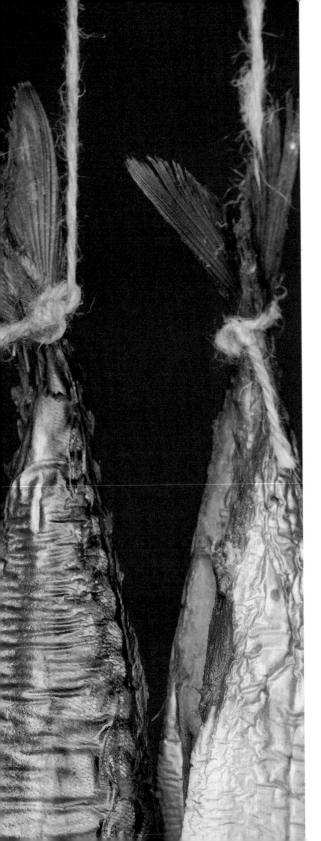

Cold smoking fish

Cold smoking is the most common method for smoking freshwater fish. Before being smoked, the fish is gutted, cleaned, rinsed thoroughly and rubbed with salt. The fish is placed in salt for a while, then rinsed and allowed to dry before being hung in a smoker.

A thin coating of proteins (called pellicle) will form on the fish during the smoking process. The pellicle kills bacteria and protects the fish. Together with the salt, this process keeps fish edible for weeks if it is kept cool and dry. Cold smoking takes hours or even days depending on the size of the fish.

Hot smoking

The process of hot smoking begins when your smoker reaches 65°C (149°F). The smoker or smokehouse needs to be airtight to reach that temperature. The method is very simple. There are various types of smoker on the market and you can buy woodchips in a plastic bag. These smokers can be laid directly on the fire and the bag of woodchips will produce the smoke.

The fish is steamed and smoked simultaneously. A small fish fillet will be ready in minutes. Larger fillets and whole fish will take more time, but seldom more than half an hour. Use the right quantity of woodchips; do not use too much. Learn by experimentation, but start with a thin layer of chips. You can make your own woodchips from juniper, oak, aspen or bog myrtle.

‹ Cold-smoked mackerel

Hot smoking fish in a closed container on a campfire. The ABU smoker is a well-known brand name for many outdoorsmen. The device shown here works on the same principle.

Salting before smoking

The fish should be salted before being smoked to add flavour and help the smoke bite better. Use less salt if you intend to hot smoke, more salt for cold smoking. The fish needs to be fresh or right out of the freezer, and it needs a good gutting, cleaning and rinse before being salted and laid in brine.

Small fish only require a simple salt rub and 30 minutes in the brine solution. Fish that weigh more than 500g (1lb 2oz) should be soaked in brine for a couple of hours; soak bigger fish overnight. Fish fillets of large salmon should soak for over 24 hours. The salt needs to penetrate into the fish down to the bone.

Salt needs more time to penetrate fatty fish such as trout or salmon, so they should always be filleted before being smoked. Remember to wash off the salt and let the fish dry properly before smoking it. Smoke bites better that way. A good tip is to rinse the fish belly before hanging or placing the fish in the smoker.

Wilderness smokers

I often encounter wilderness trenches on my journeys through forests, mountains and river valleys. They are especially numerous around the Femundsmarka National Park in Norway. Maybe the conditions here are particularly advantageous: the landscape along the riverbanks is formed by sloping morainic ridges. Nomadic people did not build smokers in haphazard places. They chose sites that they could return to, year after year. Femundsmarka has good examples of such locations. Building a smoker near camp allows you to enjoy the day's catch even more. Cold smoking gives fish a special flavour and it is a fantastic method of preservation.

A gentle slope of loose soil or sand is optimal for building a very functional smoker.

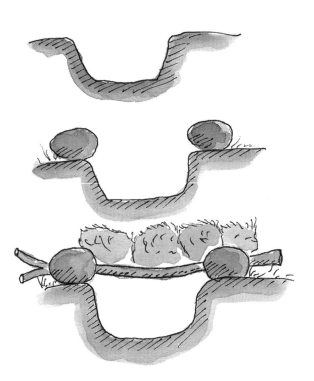

Cross-section of a wilderness smoker with the smoke duct buried below branches and turf.

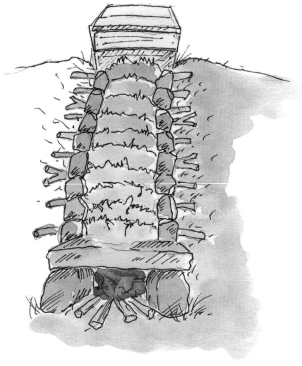

A smoker built on a slope, with the fire source at the bottom and the smoke box at the top.

104

How to build a wilderness smoker for cold smoking

A smoker consists of three parts: the fire chamber, the smoke duct and the smoke box. Choose a hillside of sand or soil, ensuring that it is not too steep for you to climb without slipping. Build a fire pit at the bottom of the slope. You can use an old metal oven if it has a door, or make an earth oven lined with stones and cover the opening with a large flat rock for adjusting airflow. Place your fire pit where you can dig a trench about 5–6m (16–20ft) long above it.

Make the trench at least 15–20cm (6–8in) deep. The trench is the smoke duct that channels smoke from the combustion chamber to the smoke box. The traditional smoke duct is a dirt trench lined with stones, but you can also use other natural materials such as wood and moss. To build a traditional smoke duct: line both sides of the duct with fist-sized rocks with no gaps between them. Lay small branches crosswise on top of the stones above the duct. Cover the branches with turf and moss so that the smoke duct is as airtight as possible.

The distance from the fire to the smoke box determines the temperature of the smoke. The distance for cold smoking should be 4m (13ft) or more; further if possible (i.e. 10m/33ft). One advantage of a traditional stone-lined duct is that soil cools down smoke rather quickly. Many people now use a concrete pipe or an old stovepipe as their smoke duct, buried below the ground.

Your smoke box is located at the top of the duct. You can build it using flat stones or flagstones, or use a wooden crate. The smoke box needs to be big enough to lay or hang fish inside. The space inside will fill completely with smoke. If the smoke is cold (-30°C/-22°F), your fish will be cold smoked.

You can start the fire as soon as the smoke box is ready. It is important to get a good fire going and to let it burn until only hot coals remain. Then lay your woodchips and sawdust on the embers. Juniper and woodchips from deciduous or fruit trees create excellent smoke. For extra flavour you can put some juniper berries or herbs on the embers.

Use branches with green leaves or live juniper sprigs to produce enough smoke. Remember, you want the sprigs and leaves to smoke, not burn. Cover any flames that appear with woodchips, juniper or other greenery.

Smoking takes time. The fire needs constant attention and must be fed with woodchips and greenery to produce smoke. The smoking process can take many hours.

I have seen homemade smokers at cabins made from an old wood stove, a metal pipe and a wooden box at the top. The smoke box is usually about the size of a beehive box – that is, about 75cm (30in) tall with walls 40cm (16in) long. The size of your smoke box depends on the length and quantity of the product you intend to smoke.

Campfire Recipes

A campfire is more than just a source of light and warmth, or a hot cup of coffee. Why not light a small fire and take a break along the trail? We all have great memories of roasting hot dogs on a stick, with sticky grilled marshmallows for dessert. Hot dogs simply taste better after a hike, roasted over flames or embers.

A frying pan or a roll of tinfoil in your rucksack is all you need to create delicious meals. Use your imagination – the only limit is how much you are willing to carry. A light meat, vegetables, some tortillas or pizza dough, and tinfoil will enable you to prepare many delightful meals over a fire. Are you an angler? Mix fresh fish with vegetables and wrap it in tinfoil. Nothing tastes better.

Stick bread

Roasting bread on a stick over a fire is great fun. You can also prepare thin baguettes on a seven-stone oven (see page 83). Stick bread takes only minutes to bake over open flame. Patience is required to get the inside baked before the crust burns! Mix bits of ham, grated cheese and herbs or spices into the dough to make it even tastier. Ham, cheese and spices will make traditional stick bread into a real meal.

Pancakes and waffles

Many people pack as little as possible for a camping trip. I like to carry a few special utensils in my rucksack to make the trip more enjoyable. I often bring a waffle iron or small frying pan. My rucksack will be heavier, but I will enjoy my time fireside even more.

Pancake or waffle batter can be prepared before leaving home and is easy to transport in a plastic bottle. If homemade batter is a challenge, why not bring a bag of shop-bought mix? It tastes great, too.

Baked potatoes

Cut a potato in half, scrape out some of the flesh and fill the space with a mix of egg, ham, and salt and pepper. Put both halves together again and pack the potato in tinfoil. This is a great way to bake a potato.

Toasted cheese bread

A packed lunch can be enjoyed as you stare into the flames of your fire. But a toasted cheese sandwich packed in tinfoil is the ultimate elevenses. It takes only minutes to get cheese to melt in your bread. Cheese bread can be prepared quickly before leaving camp. You need tinfoil, a couple of slices of bread, some butter, cheese and ham/sausage, plus a little mustard and ketchup and maybe some spices.

Spread ketchup and/or mustard on a slice of bread. Add a slice of cheese, ham and a dash of spice. Put the two slices of bread together. Spread some butter on the outside of your sandwich before packing it in tinfoil – the butter makes the crust golden and crispy. Lay the sandwich on the embers. Two minutes on each side is enough.

Fried tortillas

Corn tortillas are delicious when fried over fire. Fry some meat, chicken or fish in a pan with some spices and vegetables. Put the mix in a bowl when it is ready. Then fry your tortilla in the pan. Add a slice of cheese and fill the tortilla with some of the meat and veggies. Wrap the mix in the tortilla, then put the frying pan back on the flames. Fry the wrap on both sides, but not for too long. Try not to burn it! It is ready to eat when the cheese is melted and the tortilla is golden brown.

RECIPES

Stick bread

1.8kg (4lb) wheat flour
½ tsp salt
2 tsp sugar
2 tsp baking powder
Approximately 225ml (8 fl oz) water
5 tbsp soya oil

Mix all the dry ingredients together first, then add the water and oil. Stir everything together and knead the dough until smooth. Add a little more flour if the dough is still sticky. Put the dough in a plastic bag or container. Let your fire burn down until only a bed of hot coals remains. Twist the dough around a stick and hold it over the embers. Keep turning the stick often so that the crust is golden brown but not burnt.

Pancakes

1.3kg (3lb) wheat flour
½ tsp salt
Approximately 2 litres (75 fl oz) milk
4 eggs
Butter, for frying

To make the batter, mix the flour and salt together, then add the milk. Blend well, add the eggs, and then whisk into a batter. Let the batter rest and swell before it is fried.

Melt some butter in a frying pan. Fill the pan with a ladleful of batter and rotate the pan so that the batter coats the base in an even layer. Flip the pancake over when the surface is no longer wet. Fold the pancake in half when both sides are fried, then remove it from the pan. Stack the finished pancakes on tinfoil and keep them warm beside the fire.

Waffles

1.8kg (4lb) wheat flour
6 tbsp sugar
1 tsp ground cardamom
1 tsp baking powder
Approximately 1.75 litres (60 fl oz) milk
100g (3½ oz) melted butter
4 eggs

Mix the dry ingredients together first, then add the milk while continuing to stir. Add the melted butter and eggs. Whisk well. Let the batter rest and swell. Pour some batter into the waffle iron, place the iron on the coals, and then flip the waffle occasionally. Remove the iron from the fire. Make sure the waffle is done.

BONFIRE TRADITIONS

*Bonfires have been used
by lumberjacks, hunters
and outdoorsmen for millennia.
Fire also has many uses in religious
and cultural traditions.*

A traditional *Sankthansaften* (Saint John's Eve) bonfire with a witch effigy in Denmark

The celebration of Midsummer's Eve is the most common bonfire tradition in Scandinavia. St John's Eve is still an important ceremony in many countries to commemorate the summer solstice. In Norway and other Scandinavian countries, the tradition of gathering near a raging bonfire on Midsummer's Eve has gone on for hundreds of years.

It all began as a celebration of the summer solstice, but the celebration was Christianized as pagan lands were Christianized. The day was renamed after a Danish saint, St Johannes. For many people, Saint John's Eve is the highlight of the summer holiday season. It is celebrated with large bonfires, dancing, traditional sour cream porridge and wreaths of flowers.

Bonfires have become a much more important part of the celebration in recent times. Many communities follow strict traditions when building their bonfires. Some communities require bonfires to be built from nine different kinds of deciduous tree. Another tradition has the bonfire built around a large pole and crossbar, from which two figures are hung: an old hag and an old geezer. As the fire burns, two children are dressed up to enact a make-believe wedding as the bride and groom. The wedding is followed by dancing around the bonfire until the old hag and geezer are burned to ash. The hag and geezer represent the past. The child wedding represents the new times to come.

Some places required that the bonfire be as tall and fiery as possible. The height was important; it had to be possible to see the fire and smoke from far away, bringing tidings of the future. Lighting a bonfire often involved special rituals.

St John's bonfire, southwestern Norway

It was common to ignite the first flames without the help of matches or fire-starters, using only friction or bow drills, etc. If there was enough sun, one could use a magnifying glass. Sparks and flames created in this way were said to have magic properties.

It is easy to see why bonfires were prohibited in many countries after Christianity entered the scene. The ceremony is still full of pagan rituals.

Sweden was subject to prohibition early on. This may explain the Swedish tradition of decorating the May Pole with leaves and flowers – perhaps Sweden's most important symbol for midsummer. Swedes take midsummer very seriously. The day is an official flag day and midsummer is always celebrated on a Friday, between 19–25 June, to give Swedes a long weekend.

It is a day for families to enjoy good company, food and beverages outside. Swedes don their national costumes and join in folk dances around the Midsummer Pole.

The tradition of bonfires has gone uninterrupted in Norway and Denmark for centuries – even after both countries were Christianized. Early in the 1900s the Danes began burning the effigy of a witch on the bonfire to send her to "Blokksberg", where witches traditionally gather in Germany. New rules have been introduced for how and where local communities can light bonfires, due to fire hazards and pollution. This usually applies to areas where forest fire is a risk, and to end a long tradition of burning old rubbish and scrap in the midsummer bonfire.

Slinningsbålet (Slinningsholmen, Ålesund)

would turn a blind eye if youngsters "borrowed" materials along the way. Squads of youngsters would gather in the cities to gather fuel in the months, weeks and days before Midsummer's Eve. Competition was tough. Once the materials were collected, they had to be protected from raiding squads. It was often necessary to station guards at the site of the future bonfire to prevent competitive groups from stealing what the squad had collected.

The bonfire in Ålesund has always been quite special. The inhabitants would strive to build the most spectacular edifices, reaching higher each year, only to burn them down in one day.

In recent years the city has only allowed one large tower to be built. As compensation, it is taller than ever now. The tower is built on the same spot each year, Slinningsholmen, and constructed of barrels and industrial pallets. It has become a unique symbol and tourist attraction for the city.

Midsummer bonfire

The summer solstice is the longest day of the year, when the sun is at its highest point above the hemisphere. The sun reaches its maximum position above the horizon on 24 June. Colloquially, we say that this is the day the sun turns. The bonfire commemorates the turning. It was thought the bonfire increased the power of the sun at this critical phase and would help the sun turn and protect us from evil forces. Saint John's Eve, or Midsummer's Eve, is celebrated on 23 June.

In Norway, bonfires are generally confined to the coast. Coastal cities compete to see which community can build the largest fire. In many cities (especially in western Norway) children and young people were in charge of collecting combustible materials. People

Burning Christmas trees

Many communities in Scandinavia set aside one day a year to burn Christmas trees. This occurs on the twentieth day of Christmas. The spruce and pine trees have served their purpose and so are gathered for burning in the winter darkness on 13 January. When the fire has died, children and grown-ups are allowed to grill hot dogs on the embers.

Neighbourhoods in Germany also have a tradition of gathering old Christmas trees for a winter bonfire. However, there they save the trees until Easter. The trees are thrown into a huge pile to commemorate the end of winter and the beginning of spring.

European bonfire traditions

Midsummer bonfire traditions abound in southern Europe. Popular beliefs and superstitions speak of magic forces being released on the longest day of the year. A document penned by Martin de Arles in about 1500 describes the ritual fires used by the Basque people to ward off witches and protect crops and fields from destruction. It seems fire and midsummer bonfires have always been a popular way of warding off black magic, and have deep cultural roots throughout Europe.

Hoguera is the Spanish word for "bonfire". *La Hoguera* is celebrated throughout all of Spain at the end of June. This tradition is greatest in Catalonia and Valencia. Catalan separatists and nationalists consider 24 June to be their national day.

It has been said that the bonfire celebrations in Spain were inspired by Guy Fawkes, the British man who tried to kill King James I in London in 1605. Fawkes was a Catholic with close ties to Spain. He wanted Catholic rule across all of England. His plan was to blow up the Houses of Parliament. The Gunpowder Plot was exposed and the plan failed. Fawkes was arrested on 5 November and executed soon after. Guy Fawkes Night is still commemorated with fires and fireworks across all of the United Kingdom on 5 November each year, not unlike the Hogueras of Spain.

The Gunpowder Plot. A copperplate engraving by Crispijn Van de Passe, showing Guy Fawkes (the third man from the right).

The Smoke Signal by Fredric Remington (1905), which shows First Nation People using fire to send smoke signals.

SIGNAL FIRES

The First Nation People of North America communicated using smoke signals. Signal fires have also been common in Norway.

When a new pope is elected, the voting results are communicated by white or black smoke rising from the Sistine Chapel at the Vatican.

It is common knowledge that the First Nation People of the North American plains communicated using fire and smoke. A damp blanket was used to interrupt a column of smoke to convey messages by means of puffs that could be read at a distance as they rose to the heavens.

Smoke is still used in modern times by the papal conclave to signal the election of the new pope. The College of Cardinals informs the crowd gathered in St Peter's Square of its decision by means of smoke. An additive is mixed with the ballots as they burn to change their colour. Dark smoke signifies disagreement; white smoke means a candidate has received the two-thirds of votes that are necessary to be elected.

Beacon fires, watchtowers and lighthouses

There are many types of smoke signal. Signal fires have probably been used in Norway and other Scandinavian countries since the Bronze Age. Medieval law books in Norway describe hilltop fire towers that communicated by means of fire in times of war. The guards would alert the station on the next hill by setting their beacon on fire. This allowed them to communicate over long distances.

The Old Norse word for a hilltop signal fire (*viti*) is related to the word *vite* (meaning "to make known"). We find these words in place names around the country (such as Vettakollen in Oslo).

Signal fires in the Middle Ages

Coastal watchtowers

The entire Norwegian coastline and adjacent valleys are peppered with place names that include the words *vede, vete* or *vite*. Watchtowers along the coast were an indispensable part of the country's military defence, as we can see from this Royal Decree by Christian IV in 1628: *It is Our wish that coastal watchtowers and fire stations be arranged without delay at the most prominent and exposed locations.*

The main watchtowers formed a chain along the entire coast, making up the spine of the national signalling system. The beacon itself was made of firewood, barrels of tar and ling at the base of a pole. Long logs stood against the pole in a tepee shape.

Many hilltop stations had a small cabin made of stone nearby so that the guards could remain at their post 24/7 in both good weather and bad. The cabin had peepholes in all four walls for added security.

The last time the stations were mobilized was in 1807 during the Napoleonic Wars. The beacon at Møvig, near Kristiansand, was lit when the English naval vessel *Spencer* attacked Flekkerøy.

Lighthouses

Lighthouses have been used as navigation markers since antiquity. The Phoenicians are credited with setting up the first navigation beacons. The Lighthouse of Alexandria on the island of Pharos (in Egypt) is considered one of the Seven Wonders of the Ancient World. It was said to have been 170m (558ft) high. A fire was lit at the top of the tower at night, and by day a large mirror reflected sunlight to indicate the lighthouse's location (see page 125).

Norway's first lighthouse was erected in Lindesnes in 1655. The Lindesnes Lighthouse was established by Royal Decree on 18 July 1655. It started as a large, open fire in a metal basket, which was replaced in 1656 by a three-storey-tall wooden tower with openings facing the different sailing lanes. The first lighthouse used tallow candles. This did not provide sufficient light, so Lindesnes was shut down. It remained dark for nearly 70 years. In the meantime, a new beacon was set up in 1696 on the island of Store Færder, not far north from where the lighthouse still stands. It began as a large iron cauldron standing on the rocks. The lighthouse keeper kept it burning all night using wood and coal.

A new lighthouse was erected at Lindesnes 25 years later, still using open flame, but now inside a tower. The base of the tower had slots that provided airflow to a fire in an iron basket that was hoisted to the top of the tower. Raising the basket allowed seafarers to see the light through the openings in the tower from far away and reach land safely.

This construction was similar to other bascule lights or tipping lights, which were much bigger. They were in use in Denmark before the so-called range lights were established along the coasts of Norway and Sweden. A bascule light is simply a metal basket hanging at the end of a long pole that can be raised after the fire is lit and easily seen from sea.

The oldest bascule light with an open-basket design was built at Grenen, at Skagen Odde, Denmark, in 1627. Bascule lights were used on both sides of the Skagerrak until the end of the 1700s.

⌃ ‹ Signal fire
⌃ « Bascule light at Skagen, in Denmark
‹ Lindesnes Lighthouse, in Norway

One of the Seven Wonders of the Ancient World

The lighthouse at Pharos, off the coast of Alexandria (in Egypt), was one of the Seven Wonders of the Ancient World. It is said to have been 134m (440ft) high. The lighthouse was one of the tallest buildings of its time. Commissioned in 299 BC and built by the Greek architect Sostratos, who finished its construction in 279 BC, it was destroyed by an earthquake in 1303. The tower contained a mirror which reflected sunlight during the day and held a fire-beacon that burned at night.

The Lighthouse of Alexandria at Pharos was the tallest man-made edifice in the world for hundreds of years and is considered one of the Seven Wonders of the Ancient World. Illustration by archaeologist Hermann Thiersch (1909).

IN STOVES AND FIREPLACES

A lot of firewood is burned indoors.
In Norway, we have three million wood-burning
stoves in our homes and cabins.

Wood-burning fireplaces/stoves are either open-faced or closed-faced, and most of them contain fireplace inserts. One-fifth of all home heating in Norway uses wood as fuel. Norwegians are great consumers of firewood; we burn more than six billion kilowatt hours of it each year.

We use our fireplaces and stoves to heat one room or part of our homes. Our dependence on wood-burning technology has created an industry subject to strict requirements for the quality of the technology itself, for firewood and for optimal combustion. Recent research tells us that the most effective means of burning wood is by doing so aggressively. This means starting a fire and reloading the combustion chamber with substantial airflow. This approach provides the best heat and the most flames.

Getting your indoor fire going

To get an indoor fire going we follow the same three-step process as for firing up outside: lay a fire with tinder and kindling below thin sticks that will burn long enough to get your logs burning.

Some people swear that newspaper is the best tinder; some rely on bark, while others prefer shop-bought fireplace starters. Regardless of your choice, it is vital to keep plenty of thin kindling sticks to hand to get the logs burning – even if you use shop-bought fire-starters.

One good tip: get your kindling burning hot with plenty of air from below and keep adding sticks to keep the fire lay hot. This is true for all wood-based fireplaces, stoves and hearths. You will have to think differently, though, if you own one of the newest wood-burning stoves. Instead of building a traditional fire, the best results are achieved by filling the combustion chamber with small split sticks until it is half-full.

Ignite from the top using birch bark and kindling, or a couple of fire-starters below a thin layer of split sticks. Leave the door slightly ajar to increase airflow until the wood is burning intensely with tall flames.

A basket of bark is a good alternative to ›
shop-bought fire-starters.

I open a door of forged iron, where

split birch logs lie like red-yellow vipers

as thin green and blue flames jolt and sizzle

from the ashes.

I recall how this is

the oldest of tricks, perhaps

the oldest, for calming

young and old.

Ingvar Ambjørnsen

Advice from the experts

Researchers from The Foundation for Scientific and Industrial Research (SINTEF) are asking us to relearn old habits when it comes to lighting old indoor stoves and fireplaces. They now argue that it is best to light a fire from the top. This maximizes consumption and reduces the amount of smoke and particle emissions that escape up the chimney.

We learned as children to fire up a stove or fireplace from below, but lighting your fire from above helps burn the gases generated in firewood and increase the upward draught out the chimney. Cover the floor of the combustion chamber with firewood, then build a small pile of firewood on top of that. The small fire will eat its way downwards and heat the wood below it. Firewood releases gases when it burns. The gases burn when they reach the flames. These gases amount to more than half the energy in the wood, and it is a shame to let them escape unused up the chimney. Establishing a good upward draught in the chimney will also take longer if you light your fire from below. Manufacturers and dealers of modern stoves and inserts have done enough research to know that this is the most effective way to burn firewood when air enters from above – it is proven by the cleanliness of the glass doors, which are without soot or resin stains. A little practice lighting your fire from the top will improve combustion, and you won't need to reload as often as before.

« Lay a pile of wood in the oven.
˅ « Light from the top, and let the fire burn with the door open.
  ‹ Leave the door open until your fire burns hot.

The growth rings on denser wood are closer together, and further apart in less dense wood, with a similar difference in specific weight. This correlates to how quickly a tree grows; each growth ring tells us how much a tree has widened during a year of growth. A tree that grows slowly has denser rings, while a tree that grows quickly has broader rings.

When we talk about the specific weight of wood, we usually start with the base weight (its weight when absolutely dry) divided by its volume as unseasoned. Base weight is stated in kilos per cubic metre of wood (kg/m3 or cbm). Heat energy (i.e. calorific value) is usually calculated as kilowatts per hour (kWh) per cubic metre of wood with 17 per cent moisture content (see the table on page 135).

The types of wood with the lowest calorific values are aspen, spruce and grey alder. When purchasing so-called mixed bundles, it pays to ask what kind of wood makes up the bulk of the bundle. The best woods for indoor stoves, ovens and fireplaces are rowan, beech, oak, ash, maple and birch. Birch is the most popular, without a doubt. Birch does not provide the most heat, but it does have excellent heat energy. Birch logs come with bark, which burns easily, guaranteeing you the visual pleasure of seeing flames. Pine is nearly as high in energy as birch. Pine flames are lighter in colour and well suited to enclosed fireplaces with large glass fronts. Some argue that burning pine increases the risk of chimney fire due to its high creosote content, but the researchers at SINTEF argue that pine is not riskier than any other type of wood when it is burned correctly.

Use dry firewood

Dry firewood is essential – the moisture content of dry wood is below 20 per cent. Moist wood burns poorly; the energy in moist firewood is used to expel the moisture instead of burning the wood. You can test the dryness of wood using sight and sound. Look for cracks on the ends of a log. This is a sign of dryness. Hit two logs together. If you hear a dry clink, then your wood is dry. If you hear a low clunk, the wood is moist.

Wood energy and calorific value

The calorific value of wood varies depending on its specific weight and moisture content. Specific weight varies greatly from wood to wood, but the rule-of-thumb is that trees growing in denser forests have a higher density and often a higher specific weight.

Birch

Dry birch has a lovely, slow and stable burn. Stacked beside your fireplace, it is very attractive. It smells good, too. The bark will get the chamber filled with birch logs burning fast. It also leaves a warm bed of embers in the chamber when the flames have died.

Spruce

Some kinds of spruce burn quickly, while some burn slowly. It depends on how quickly the tree grew. Spruce logs with broad growth rings burn faster with less heat. Older, slow-growth spruce trees are denser; the growth rings are narrower. This type of spruce is loaded with potential energy. It burns effectively and hot – but can spark and crackle, so you need to keep the door closed. Spruce makes the ideal kindling. It is easy to split, ignites easily and is an excellent catalyst for burning larger logs.

Pine

Dry pine makes good kindling, and larger logs have good heat energy. Pine burns with tall bright flames. That makes it excellent for fireplaces and ovens. Pine is also a good choice for ovens with large glass fronts. The bright flames light up the whole room.

Oak

Oak has a very high calorific value. It burns long and hot once you get it burning. Oak needs to dry thoroughly due to its high density.

Maple

Maple is almost as dense as oak. It would be a very popular source of firewood if it were more abundant. It is dense and burns slow and hot.

Aspen

Many people think of aspen as the wood that matches are made from. Aspen has a low calorific value and burns with calm flames. Aspen is just as easy to split into thin straight sticks as spruce. This makes aspen a suitable choice for kindling and for fire lays in smaller ovens, such as kitchen stoves and small baking ovens.

Grey alder and common alder

Grey alder is low-density and grows quickly. Its calorific value is lower than that of aspen. Common alder, however, is significantly denser and has more heat energy, which is almost equal to that of pine.

Willow

Willow is similar to common alder, with high density and good heat energy.

Ash

Ash is slightly denser than birch and almost as dense as oak. This makes it good firewood. Strangely, ash burns even when unseasoned.

Heat Energy
Specific weight and calorific value

Type of wood	Specific weight kg/cbm	kWh per cbm
Holly	675	3496
Yew	600	3107
Beech	570	2963
Ash	550	2859
Oak	550	2859
Elm	540	2807
Rowan	520	2703
Larch	540	2796
Maple	530	2755
Hazel	510	2641
Birch	500	2589
Bird cherry	490	2538
Pine	440	2287
Alder	440	2287
Lime/basswood	430	2235
Willow	430	2235
Aspen	400	2079
Spruce	380	1975
Grey alder	360	1864

(Source: *Norsk Ved*)

ALMIGHTY FIRE

*Man discovered the power of fire
hundreds of thousands of years ago.
We learned that it could be used for more
than just light and warmth, and for cooking.*

Fire allowed humans to
control light and temperature
– yes, even life itself.
Fire changed everything.
It changed our food habits; humans became
stronger and more aware of
the significance of strategic choices.
Humans began using fire for
cooking and roasting foodstuffs, hardening
wood and burning clay.
As time progressed, man learned to melt
and fashion metal into tools
and more effective weapons.
Yet we never
lost respect for FIRE
– the almighty flames and
the intense heat it provides.

Fire as catalyst

We learned about the catalytic properties of fire as soon as we discovered it. We learned that water vaporizes when it hits fire and that water turns to steam in a pot. We learned how to burn clay by picking through the ashes of a dead campfire. We learned that certain minerals release metals when heated; some even turn liquid. Silica sand and ashes turn into glass when heated. The people who mastered these mysteries were valued for their skills.

Witch-meal

You can make magic explosive flames using a powder found in nature. This fantastic substance is called witch-meal, which is the colloquial name for the pollen from club moss (*Lycopodium*). *Lycopodium clavatum* is the species with the most witch-meal. Club moss sporangia form in early October. To make witch-meal, you need to pick the sporangium and dry it on a plate in a warm place overnight. Shake the plate, and the dry sporangium will release a fine yellow powder.

Witch-meal powder burns fast, spreading through the air above the flame due to its oil content (50 per cent plus) and consuming oxygen as it spreads. The powder will not ignite while on the plate. I used to buy witch-meal at the pharmacy when I was young, but it has only been available online in recent years (www. vitenwahl.no).

Lycopodium annotinum (witch-meal) ›

141

Blast furnace (Bærums Verk, in Norway)

Primitive ceramic oven. Clay jars are converted to ceramic when they are fired in a kiln at high temperature.

New technology

Every petrol engine relies on a spark to ignite a flammable substance to release energy and create an explosion. This is one example of technological development through the mastery of fire.

Potters and blacksmiths

Clay heated to a temperature of between 700°C (1,292°F) and 1400°C (2,552°F) undergoes a chemical and physical change that makes it hard, strong and insoluble in water. The first ceramics were simple clays roasted in a fire pit. Ceramics remain one of the least expensive and most sold products in the world. Bricks, cooking pots, *objets d'art*, plates and even musical instruments can be made from baked clay.

A technique called *cire perdue* (which means "lost wax") became popular among blacksmiths in the Bronze Age. The artisan starts by creating a model of an object in clay. The object is covered in beeswax and then packed in a second layer of clay. The form is then burnt in an oven so that the beeswax melts, leaving a baked mould. Melted bronze is then poured into the empty space left by the wax. The clay is then removed to expose the bronze object.

Bloomery furnaces and smelting iron

The art of extracting iron from iron ore has existed in this part of the world since the Pre-Roman Iron Age (approximately 500 BC) until the end of the 1700s. Clumps of so-called swamp ore were harvested from bogs and swampland. The raw ore was put in an open fire pit to reduce the iron ore to its iron oxides.

Metallurgy in the Middle Ages

The iron oxide was burned in a bloomery fired with wood-charcoal. The walls of the furnace were often made of clay. The clay walls had to be broken up after each firing to remove the bloom (iron and slag). The bloomery's design did not change much through history, but they did get larger in the fifteenth century and were in use until around the mid-1800s. They were often constructed below ground in shaft furnaces with an air source provided by two large bellows. The slag had to be removed before the iron could be used to forge tools and weapons. Wood-charcoal remained the fuel source throughout this time.

Tar kilns

Tar is distilled by heating resinous trees (a process known as dry distillation). Tar is historically made from resinous wood. Resin is the fatty substance found in heartwood and in the roots of pine trees. Tar kilns are the most common method of extraction. Scandinavians have been producing tar since the Viking Age, about a thousand years ago.

The bottom of the kiln is sloped to allow the tar to trickle out. The kiln is usually built on a slope and lined with stones or heavy logs. Then it is covered with a layer of birch bark. The bark is dense and stacked so that hot tar trickles down the funnel into a channel made from a hollowed-out log, before finally dripping into a barrel or container. The biggest tar kilns could hold as many as 1,500 pieces of bark.

The roots of pine trees were traditionally used to distil tar in Scandinavia. There is a lot of resin in the roots of pine trees. The best roots are found in the old stumps on which the outer layer of wood has rotted away. Pine stumps need to be 10 years old, or older, and it was common to use stumps that were hundreds of years old. The work of stubbing (uprooting old stumps) began in the autumn after the ground had thawed. The material was transported to the kiln to be cleaned and chopped into blocks before use.

143

Making tar in a kitchen pot

Anyone with a few pine trees in the garden could make tar in a large pot at home. Thin sticks were placed upright on the bottom of a pot – densely packed side by side. The last stick was hammered in, to fill the whole pot. A flat stone was made ready beside the fire. The stone had a small groove in it. The pot was placed on the stone on its side, and the edges of the pot were covered with soil or clay to stop airflow. A fire was built over the pot. The tar began to drip onto the stone and run down the groove into a catch basin. The pot was stood upright again when the tar stopped dripping. What remained inside was pure charcoal.

You can try this at home:

You need two metal containers and a piece of sheet metal with a hole in the middle. Bury the first container in the ground with the opening facing up. Then cover the container with a metal sheet and press inward, creating a simple funnel. Fill the second container with finely split fatwood. Put the second container upside down above the tap hole. Seal the edges with moist dirt. Build a log-cabin fire around the container and light from the top. The fire will have died after 30 minutes or so. The fatwood in the second container will have turned to charcoal and the other container will hold a little tar.

Tar kiln, Finland, in approximately 1900

Tar kettle, Åstdal Days Festival, Ringsaker, Norway

The wood was then split into sticks, sorted and stacked to dry. The tar kiln was built around a large log or pole, which acted as a reference point for building a round symmetrical kiln. The kiln became smaller as it grew, so creating the traditional round igloo shape. The outer layer was covered with woodchips, and the pole was removed to leave a hole at the top and a hole at the bottom where the funnel led to the channel. Then the whole kiln was covered with turf and moss, with the soil side facing out to insulate against wind. An opening remained at the bottom. This was where the kiln was lit. The tar could be ruined if the turf/moss cover started to burn.

The resinous wood should not burn. It needs to get hot with a minimal supply of air in the kiln. This allows the tar to melt and drain from the wood. The secret is keeping the temperature at about 40°C (104°F). It takes three to five hours before the first drops of tar appear. The first bucket is dilute, containing lots of water and turpentine. The tar gets darker and thicker as the hours pass. A large tar kiln will produce tar for several days. When the kiln stops producing, it is covered and soaked with water to extract the final drops. A large quantity of charcoal remains when distillation is complete.

The Flesberg charcoal pile supplied the Vinoren Silverworks with charcoal until 1878.

Charcoal piles

Charcoal is made in a charcoal pile. Making charcoal requires incomplete combustion. This is achieved by controlling the amount of air entering the pile to convert unseasoned wood to charcoal. This produces a fuel source that is virtually without water. The absence of moisture in the charcoal allows it to burn at a higher temperature than dry firewood. This was especially important in the process of extracting iron from iron ore.

The traditional charcoal pile did not change until the end of the 1800s. We can find remnants of old charcoal piles anywhere that we find forests. A charcoal-pile site is easily identified as a large circular area with a diameter of up to 20m (66ft), with no vegetation, surrounded by a mound of dirt, and lots of ash on the ground. There are many place names in Norway that refer to such sites: Kølabånn, Kullbunn, Persbonn and Kolbotn in the Oppegård.

Charcoal piles are built on flat ground. A stake – the quandel – is driven into the ground, to become the heart of the pile. The quandel is braced if necessary for support. Logs are then laid on the ground so they fan out from the quandary to the outer limit of the pile. Spilt logs are laid on top of these, so creating air pockets between them and the ground. When the floor is completed, the charcoal burner begins to stand dry pieces of wood against the quandary to create a shaft at the "heart" of the pile. The shaft will be used to light the fire and create temporary airflow inside.

Wood-charcoal is still in use by blacksmiths and for outdoor barbecues.

More spruce and fatwood is piled at the bottom, next to the quandary. This is done to create an extra hot core that will stay hot throughout the charcoal production process. The wood is stacked outward, still in an upright position and leaning inward towards the heart. Pieces of wood of varying sizes form the outer layer to make the surface as flat as possible before the whole pile is covered with green spruce branches and turf. Finally, the whole pile is covered with a 20-cm (8-in) thick layer of clay soil. Small channels are left open at the bottom of the pile to supply oxygen for combustion.

The process starts with lighting a fire at the top of the heart. Once the fire is burning it is pushed into the vertical channel (the heart) where it ignites the dry spruce and fatwood at the base. The intense heat created at the core spreads to the rest of the wood, without igniting into flames, and so creates charcoal. Large charcoal piles are tended by the charcoal-burner who adjusts airflow. It can take weeks for a pile to convert all the wood to charcoal.

Ironworks in the region were the largest consumers of charcoal. We find the biggest ironworks in Norway, where the raw materials for making charcoal were most abundant. The need for charcoal diminished rapidly after 1860 when most ironworks were either decommissioned or began using coke as fuel. Wood-charcoal was also used by local blacksmiths, so the tradition stayed alive until about 1900.

Fire-setting

Fire-setting is a method of mining rock and was in use from prehistoric times to the 1800s, until gunpowder and dynamite were invented. The principle was simple; a large controlled fire was started beside the rock face to be mined so that the flames heated the rock. The firewood for fire-setting was comprised of 1-m (3-ft) long logs split to about 15cm (½ft) wide, using logs of spruce and/or pine. The hot rock was cooled rapidly by dousing it with water. This caused the rock to expand and fracture through thermal shock, often breaking up into shards or small pieces. The workers waited until the smoke cleared and the rock had cooled before chipping away at the rock face using various tools. Workers would progress into the rock face at a rate of 1–3m (3–10ft) a month – a slow and resource-intensive process that required much wood.

Though it appears simple, mining and fire-setting required expert knowledge and experience. Not all rock types respond to fire and heat, and the firewood chosen depended on the type of rock to be mined. Another important skill that required know-how was building a fire to get it to burn just where you wanted it to burn. Firewood could not be wasted. High flames and intense heat were most important, not how long the fire burned for.

Split wood was preferred because it had those properties. The objective was to use less wood to generate more heat, fast. A long stick was used to light the fire. Pine was preferred. The end of the stick was feathered so that the shavings stuck out like a beard.

Fire-setting, a copperplate engraving by Gregorius Agricola

Documents from 1784 about fire-setting from the Kongsberg Silver Mines tell us:

The main fire was built of logs with a small fire below it made of split wood. Loose rock was knocked out once the fire had died. The burnt stone was cleared out and what remained of firewood and ashes was raked against the rock face for the next burn.

Spruce is the best fuel; it burns with … *intense rapid heat without leaving many embers, which hinder the workers' progress.*

Fire creates shear stress in rock, making it brittle and causing it to break into shards. This picture shows a fire-setting site in a pothole near Lyngør, in Norway.

WHEN YOUR FIRE IS BUT EMBERS

Remember to put out your campfire before moving on; do this right or it might flare up again.

A campfire leaves its mark on nature, as is shown here.

Your campfire is not dead until you can put a flat hand on it and feel no warmth.

Nils Bloch-Hoell,
Norwegian Trekking Association

When your campfire has died out and it is time to go, it is very important to extinguish it and clear the site before heading off. The best way to extinguish warm embers is by dousing them with water. If you do not have enough water to hand, do not leave until the embers are cold. Spread out the embers to speed up the process. The final embers can be extinguished by removing their supply of air. Cover everything with sand and dirt.

Clear up the fire site

Ethical campers try to live up to the motto: *leave no trace*. When the fire has died, try to eliminate any signs of your fire site and your presence, as much as possible. Get rid of the ring of stones you made and disperse any firewood you may have left – leave only the charred earth of your fire. If possible, let your fire burn out completely so there is no charcoal or half-burnt branches left; it will take decades for the wood to decay. This approach also ensures that no warm embers remain.

Leave no trace

Everyone who visits forests, fields, mountains or beaches should plan to take everything they carried in to nature out with them when heading home. Some waste can be burned on your campfire, but most modern packaging will not burn completely.

Most campfires are lit in places to which we never intend to return. With that in mind, remember to hide all traces of your visit; Mother Nature does not need to be scarred by the remains of campers.

Permanent campsites and fire pits

The situation is a little different when visiting permanent camps, but that does not relieve us of the obligation to clean up before leaving. The idea is to leave the campsite perfect for the next visitors. Campers enjoy arriving at campsites where the rubbish has been removed and the firewood is stacked and waiting. Arriving at a site where the previous visitors showed concern for others is a nice feeling, and it can motivate us to do the same.

A good habit to get into: clear away your fire site before moving on. ›

153

Campfire under the stars

We are fascinated by the contrast between the darkness and the light of dancing flames. The campfire has always been a place for fellowship. Fire provides light, heat and security, and meals can be prepared over the flames. Perhaps it was the sum of all these very human things that piqued my interest in outdoor life and my attraction to hiking. I bring my nature experience home with me; the smell of smoke in my clothes, hair and beard.

That is how it should be.

Literary Sources

Page 16: Gisle Skeie: from his hymn entitled *Gud gav oss ilden*, 2007 (www.nyesalmer.no)

Page 19: Hans Børli: "Ilden", *Dagsavisen* (newspaper), *Glåmdalen*, 29 July 1961, taken from *Samlede dikt*, Aschehoug, 1995

Page 26: Gil Adamson: *The Outlander*, 2007. Norwegian version: *Enken* (the Widow), Forlaget Oktober, 2011. Norwegian title: *Enken*, translated by Kia Halling

Page 32: Rudolf Nilsen: "Ved bålet", from the collection of poems entitled *Hverdagen* (Everyday), Gyldendal Norsk Forlag, 1929. Also published in *Samlede dikt*, Gyldendal Norsk Forlag, 1946

Page 37: Jack London: *To Build a Fire*, published in Norwegian in *Livsgnisten og andre noveller*, Den Norske Bokklubben, 1976. Translated into Norwegian by Kjell Risvik

Page 40: Anders Larsson-Lussi, from Yngve Ryd: *Eld*, Natur & Kultur, Stockholm, 2005

Page 49: Nils-Henrik Gunnare, from Yngve Ryd: *Eld*, Natur & Kultur, Stockholm, 2005

Page 51: Jack London: *To Build a Fire*, published in Norwegian in *Livsgnisten og andre noveller*, Den Norske Bokklubben, 1976. Translated into Norwegian by Kjell Risvik

Page 63: Isak Parfa, from Yngve Ryd: *Eld*, Natur & Kultur, Stockholm, 2005

Page 66: Hans Børli: "Tyrielden", *Dagsavisen Kongsvinger Arbeiderblad*, 1 February 1942, taken from *Samlede dikt*, Aschehoug, 1995

Page 91: Troels Lund: from the book *Daily Life in Scandinavia in the 16th Century*, Gyldendalske Boghandel, Copenhagen, 1914

Page 130: Ingvar Ambjørnsen: *Opp Oridongo*, Cappelen Damm, 2009

Page 152: Nils Bloch-Hoell, Norwegian Trekking Association, published online (www.ut.no.)

Other Sources

Ottar (Norwegian popular science magazine, no. 262):
"Ilden i sentrum", Tromsø Museum, 2006
Yngve Ryd: *Eld – flammor och glöd – samisk eldkonst.*
Natur & Kultur, Stockholm, 2005
Lars Mytting: *Hel Ved*, Kagge forlag, 2011
David More: *Trær i Norge og Europa*, Cappelen Damm, 2005
Bjarne Lindbekk: *Våre skogstrær*, Omega, 2000
Bjarne Lindbekk; *Treet, skogen og mennesket*, Omega, 2006
Jakt, fiske og friluftsliv i Norge, Kunnskapsforlaget, 2005
Stefan Källman and Harry Sepp: *Overleve på naturens vilkår*, N.W. Damm & Søn, 2001
Lars Monsen: *Villmarksboka*, Lars Monsen
Boksenteret Outdoors, 2005
Lars Monsen: *101 Villmarkstips*, revised edition, Larsforlaget, 2013
Øivind Berg: *Barnas bok om friluftsliv*, N.W. Damm & Søn, 1992
Espen Farstad and Dag Heyerdal Larsen: *Uteboka*, Gyldendal Norsk Forlag, 2000
Anne B. Bull-Gundersen: *Året rundt*, Aschehoug, 2001
Øivind Berg: *Vilt og tøft – Villmarksboka*, N.W. Damm & Søn, 2000
Øivind Berg: *Ut på tur – friluftsliv*, N.W. Damm & Søn, 2002
Øivind Berg: *Kom ut – barnas store bok om friluftsliv*, N.W. Damm & Søn, 2002
Norges Historie, Aschehoug, 1994
Olaus Magnus: *Historia om de nordiska folken*, Gidlund, Stockholm, 2010
Store norske leksikon, Kunnskapsforlaget, 2005
Arne Johan Gjermundsen: *Det gamle verk – Ulefoss Jernbværk rundt år 1800*, Norgesforlaget, 2007
By og Bygd, Norsk Folkemuseum, 1972
Norwegian magazine: *Norsk ved*

Credits

THIS IS A CARLTON BOOK

This edition first published in Great Britain in 2017 by Carlton Books
An imprint of the Carlton Publishing Group
20 Mortimer Street
London W1T 3JW

© CAPPELEN DAMM AS 2015

A CIP catalogue for this book is available from the British Library.

The author has received economic support from the Norwegian
Non-Fiction Writers and Translators Organization to write the
manuscript for this book.

Translated by Jeffrey Engberg
This translation has been published with the financial support of
NORLA
Illustrations and photographs by Øivind Berg, if not otherwise stated
Interior design by Erlend Askhov
Cover design by Natasha Le Coultre

ISBN: 978 1 78097 991 5

Printed in China

10 9 8 7 6 5 4 3 2 1

This is a book about making fires in the context of camping and
cooking outdoors, and the information contained in it is not to
be used for any other purpose. However, the instructions in this
book are for general information only and the publishers take no
responsibility for any consequence of you using the information
contained herein, especially if you harm yourself accidentally.

Also available:

978-1-78739-003-4